Praise for Stephen Baxter's

Ages in Chaos

"Baxter, best known for his science fiction, does a commendable job of contextualizing Hutton's career within the Enlightenment."

—*Publishers Weekly*

"Hutton more or less invented the earth sciences and the modern understanding of geology and laid the foundations for Darwin's work over the next few decades. . . . Baxter takes an unwavering and mostly sympathetic look at his life and ideas."

—*Focus*

"Baxter, a leading science fiction writer, employs his storytelling skills to lucid effect in a highly readable biography of a fascinating, neglected figure."

—*Time Out*

Also from Tom Doherty Associates

The Light of Other Days (with Arthur C. Clarke)

Ages in Chaos

*James Hutton and the
Discovery of Deep Time*

Stephen Baxter

A Tom Doherty Associates Book
New York

AGES IN CHAOS: JAMES HUTTON AND THE
DISCOVERY OF DEEP TIME

Copyright © 2003 by Stephen Baxter

Originally published in 2003 by Weidenfeld & Nicolson, Great Britain,
under the title *Revolutions in the Earth*.

Edited by David G. Hartwell.

A Forge Book
Published by Tom Doherty Associates, LLC
175 Fifth Avenue
New York, NY 10010

www.tor.com

Forge® is a registered trademark of Tom Doherty Associates, LLC.

Library of Congress Cataloging-in-Publication Data

Baxter, Stephen.
    Ages in chaos : James Hutton and the discovery of deep time / Stephen
Baxter.
        p. cm.
    Includes bibliographical references (p.[233]-237) and index.
    "A Tom Doherty Associates book."
    ISBN-13: 978-0-765-31268-6
    ISBN-10: 0-765-31268-9
    1. Hutton, James, 1726–1797. 2. Hutton, James, 1726–1797—Friends and
associates. 3. Geologists—Scotland—Biography. 4. Geology—Scotland—
History—18th century. 5. Scotland—Intellectual life—18th century.
I. Title.

QE22.H9 B39 2004
550'.92—dc22
[B]
                                                                    2004061940

First Hardcover Edition: November 2004
First Trade Paperback Edition: August 2006

Printed in the United States of America

0 9 8 7 6 5 4 3 2 1

To my mother

CONTENTS

Frontispiece: Portrait of James Hutton by Sir Henry Raeburn, *The Scottish National Portrait Gallery*

Siccar Point: Author's photograph

Siccar Point

PROLOGUE

On a bright summer day in 1788, James Hutton and two close friends took a boat trip to Siccar Point, a sea cliff on Scotland's eastern coast. Their purpose was geological investigation.

In April 2002 I made my own pilgrimage to Siccar Point. My wife and I had hired a cottage on the outskirts of Dunbar, some thirty kilometres east of Edinburgh. Now we drove east along the A1, the trunk road that leads to Berwick and England, turning off along an A-road that hugs the shoreline.

This north Berwickshire coast is spectacular. A glance offshore reveals the uncompromising plug that is Bass Rock, the core of an ancient volcano. More plugs protrude from the ground in a great line that marches to the west through Edinburgh and beyond. It is a stitching of volcanism that marks the place where Scotland collided with England during the building of a supercontinent, hundreds of millions of years ago. If the rocks show a turbulent geological history, then the battered castles of the area betray much human turmoil. Today Scotland is at peace, the castles monuments for tourists taking time out from the golf courses. You can take a boat trip around Bass Rock. Its upper surface is made white by the bodies of sea birds, puffins and gulls and eider ducks and gannets, a hundred thousand of them.

We took a couple of turns along minor roads, passing the ruins of a Saxon church. We came to a factory nestling in an old quarry, where we parked the car. We scrambled up to the cliff top. The land was green and tumbled, littered with yellow gorse and dry stone walls. The cliffs were steep, the rocks distorted, their geology complex even at first glance. Cutting across cattle fields, we walked east beyond the Point itself, until we could get a good view back. We saw beaches of rock and sand, and the grey-blue bulk of Torness. The nuclear power station dominates the scenery in our era, but the geology is still there.

The beauty of the Point is that here the sea has worn away the land's green cover, exposing the rocks beneath. You can easily see a contact between two types of rock: red sandstone strata overlie darker and older 'greywacke' (a hard grey sedimentary rock). The red sandstone is pretty much horizontal, but the underlying greywacke strata are tilted up, in places almost vertically, and they stick through the overlaid sandstone like broken teeth.

It was this exposure that Hutton had come looking for. He believed that the subtle features of this 'unconformity' were proof of the extraordinary idea that had obsessed him for nearly thirty years: that the Earth was not a mere few thousand years old, as Biblical scholars insisted, but vastly ancient, and that every particle of its surface had undergone countless cycles of erosion and uplift.

Standing on that windswept cliff top, we tried to imagine how James Hutton and his friends came upon this place, more than two centuries ago. Hutton was already sixty-two. His companions that day – the chemist Sir James Hall and mathematician John Playfair, both much younger, like Hutton both Edinburgh gentlemen – scrambled clumsily but gamely after Hutton onto the shore. They listened loyally as Hutton, intense and verbose, lectured them about the features of the Point.

We must read the rocks, said Hutton, who had taught himself how.

Strata of sedimentary rocks are laid down in oceans, and they are always formed horizontally. They are the products of erosion, the everyday wearing away of rock by air or water, the removal of just a grain at a time, washed down from the land to the sea floor. Just a grain at a time – but given *enough* time, that ceaseless chiselling forms great layers of sea-bottom debris, which are consolidated by heat and pressure and chemical cementing into rock.

Sometimes such deposits can be formed, metre after metre, year after year, with no significant change. On the coasts of southern England there are chalk beds hundreds of metres thick, laid down out of a sea that was warm when dinosaurs chased across the land. More usually the process of formation of sedimentary rocks is interrupted. These events may be subtle, just slight changes in environmental conditions which give rise to 'bedding planes', as the geologists say, the markers that delineate for us the rocks' familiar layers of strata.

The breaks may be more spectacular, however. Great forces within the Earth can distort the neat layers. The strata can be lifted or lowered, tilted up or down, broken and folded over, solid rock moulded like putty. And when the rocks have been lifted away from their formative sea beds and into the air, they are immediately subject to the same relentless forces of erosion which led to their creation in the first place – and somewhere new strata are laid down from the debris.

It is this combination of distortion, erosion and fresh deposition that can give rise to the geological record's most spectacular discontinuities of all: the unconformities. An unconformity is a great gap in time, where rocks separated by millions of years are jammed into contact.

This is what Hutton recognised at Siccar Point, the daddy of all British unconformities. After the formation of these greywacke strata – so Hutton explained – geological violence followed. The land here was crumpled and raised up: the greywacke strata were broken and tilted, and became part of a great mountain range, called the Caledonian. (We know now that this was a side effect of Scotland's collision with England. When continents collide mountains are formed, just as the Himalayas today are the result of the driving of India into Asia.) Immediately erosion began its relentless work. Again, of course, the erosion was slow, just a grain at a time, almost imperceptible to our eyes, or even over a human lifetime. But given enough time the mountains were worn to nubs, and where they were exposed to the air the tilted greywacke strata were *sliced through*, as neatly as if a white-hot wire had been passed through ice cream. You can see such levelling all along the Berwickshire beaches close to Siccar. All it took was time.

Still the geological processing continued. Next, in a new tectonic spasm, this levelled beach was lowered beneath a new ocean – or perhaps the ocean rose to cover it. Debris washed from the dry land began to settle on the submerged surface – again, just a grain at a time, but after enough time more flat strata formed, laid down over the sliced-off edges of the older rocks.

Thus the Siccar Point unconformity was formed, a junction between rocks separated in their creation by eighty million years. Erosion, deposition, consolidation, uplift, over and over again, each new landscape laid down on the ruin of the old: the Point's jumbled strata are nothing less than the wreckage of worlds – and a record of their creation and destruction.

Previous generations had imagined that it had taken cataclysmic, probably divine, interventions to create such spectacular formations: Noah's Flood, for example. Hutton's genius was to recognise that it was not a miracle that was required, that the processes of erosion and uplift that we see at work around us today are sufficient to explain everything in the rocks – sufficient, *if* they are given long enough.

Hutton replaced the hand of God with the great pressure of time, long aeons of it. And deep geological time is Hutton's wonderful and terrible legacy, to us and every generation to come.

You can see how Hutton looked in Sir Henry Raeburn's portrait of him (see Frontispiece). The picture hangs in the Scottish National Portrait Gallery in Edinburgh, in the section marked 'In the Age of Burns and Scott'. Hutton is sitting in his study, surrounded by the insignia of his interests: the manuscript of his great book *Theory of the Earth*, a quill pen – and a few rocks, just small samples, some of the precious 'hand specimens' he used to make his geological points, such as a bit of chalk bearing the impression of a fossil sea shell.

The picture is an early Raeburn and not a great one. Though Hutton seems informal, with the lower buttons of his brown waistcoat undone, he looks uncomfortable, with his legs crossed, one arm draped over the back of his chair, and his hands clasped together. He has a thin face with a prominent nose and a weak chin, and wisps of grey hair surround his bald scalp. He is looking away from the viewer, as if nervously. He was painted when aged about sixty. For a science buff like me it is good to see a geologist so honoured in a hall named after poets, even if his portrait is overwhelmed by the huge marble statue of his friend James Watt in the Gallery's Great Hall. But for a great man of Edinburgh's Enlightenment Hutton looks oddly vulnerable: perhaps he was a man of contrasts, as most of us are.

I first became interested in Hutton when I visited Edinburgh in 1997, to research a science fiction novel I planned to set there (a geology-based disaster story in which I took gleeful pleasure in destroying Torness). Much of Edinburgh's beauty comes from its geological setting. Edinburgh's castle is built on top of one of those supercontinental volcanic plugs – the stones are laid directly onto the basalt, like a fantasy palace from Peake or Tolkien. Tens of thousands of years ago an ice sheet moved east over the site of Edinburgh, a monstrous slab so thick that even Arthur's Seat was a small obstacle in its path. But the irreducibly stubborn volcanic cores gave some shelter to the softer rocks around them, and in the lee of the plugs great tails of rock remained, like shadows. The spine of Castle Hill's glacial tail has been built over, to become the High Street and Canongate, today cluttered with tourist-trap shops.

In Edinburgh, you are surrounded by time, deep time frozen in the rocks on which the city has been heaped. Compared to such antiquity, human time is fleeting.

We struggle to map our own lives, marking unravelling years with birthdays and anniversaries, grasping at fading memories with mementos and photographs, with gravestone epitaphs and monuments and Internet genealogies. But a human lifespan is brief indeed. Over longer timescales, nothing is constant. James Hutton was born nearly three centuries ago. Over such intervals even languages evolve: basic vocabularies change by around twenty per cent over a thousand years. Hutton's own letters take some decoding today. We can still read Shakespeare, but need notes to help figure out some of his archaisms. Many of us might struggle through Chaucer, but *Beowulf* is beyond anyone but the experts. Nations change too, of course. When Hutton was born, there was no United States: much of North America was still a British colony. Still, some nations have survived to the present from Hutton's time – Britain itself, for example.

Look back only a little deeper, say to a thousand years, and pre-Conquest British history dissolves into turbulence and mist. Virtually no political institutions survive over a millennium, though religious organisations persist, such as the Catholic Church. A thousand years ago most Europeans scratched out lives in crude villages; the advanced cultures of the world were in the east – the Arabs, the Chinese.

Add another zero, look back *tens* of thousands of years, and you have already reached the limits of humanity. Ice Age cave paintings speak to us from a time before anything like modern civilisation had been born, but we will probably never understand their message. It may be that modern humanity itself is not much older than the earliest of our artworks.

But a few tens of millennia, the age of humanity itself, are only the surface of the great ocean of past time. The first recognisable hominid tools – bits of chipped stone – are around *a hundred times* older than the first cave paintings. Go back that far and the ancestors of humanity were like upright chimps. Another factor of ten, and there were no apes at all. The most advanced primates were the size of rats or squirrels – but they survived the impact of a comet upon the Earth, a trauma whose scars can still be seen in the rocks, in a layer of compacted ash and dust.

Even the monstrous interval since the death of the dinosaurs is only a little more than a hundredth of the true age of the Earth. The layers of rocks *beneath* the comet ash tell the story of aeons of chthonic churning, a great mindless time that preceded the arrival of anything even remotely like ourselves on the planet.

And, given enough time, even the land changes. 'The shepherd', Hutton wrote, 'thinks the mountain, on which he feeds his flock, to have always been there.' But look into time's deepest abyss and even rock flows like water, mountains billow like clouds, and continents spin in a basalt sea.

This is our modern view of time. The late Stephen Jay Gould called it 'geology's most frightening fact', because 'if humanity arose just yesterday as a small branch on a flourishing tree, then life may not in a genuine sense exist for us or because of us. Perhaps we are only an afterthought, a kind of cosmic accident ...'

We didn't always think this way.

When James Hutton was born, in the first half of the eighteenth century, most educated westerners thought the world was just six thousand years old, as computed according to the Bible's chronology. *Six thousand years*: just a few hundred human generations, a planet not much older than the pyramids. It was a comforting vision of a young Earth ruled by mankind since its origin.

The brevity of time was one of three struts that reassured mankind of our special nature: our world was at the centre of the universe, the world had been created by God for the purpose of supporting us, and the universe was only a little older than we were – in fact only *days* older. Today all the legs of that great tripod of certainty have been knocked away. Copernicus showed that the Earth, far from being the centre of things, is just a mote adrift in a great sea of darkness. Darwin would show that we are not a divine creation but the product of selection and adaptation: slow, orderly but mindless processes. And James Hutton proved that the Earth is not as young as mankind, but vastly older. As early as 1788, this Scottish amateur scientist perceived revolutions in the Earth, and declared that the geological record revealed 'no vestige of a beginning, no prospect of an end'.

The demonstration that we are as lost in time as in Copernicus' space has surely been the most extraordinary upheaval of all in modern human thinking – as well as the most essential, for without it the work of Darwin would have had no context; there would have been no time for evolution to do its work. But today hardly anyone knows Hutton's name.

Who was Hutton? How could a gentleman scientist-philosopher of the late eighteenth century, without even a formal academic post, come up with such a startling and modern view of the Earth?

The received view of Hutton, I found, was unsatisfactory. Donald McIntyre and Alan McKirdy published a brief popular biography of Hutton in 1997. This engaging book, full of pictures, describes the geology well – but Hutton, bachelor scholar in Enlightenment Edinburgh, sounds like an ascetic paragon of scientific virtue. The *Encyclopaedia Britannica* sketches a modern-sounding observational scientist who studied the rocks, and thereby showed that the world's geological phenomena could be explained by processes we can see today: 'From that time on, geology became a science ... No biblical explanations were necessary.' In such accounts Hutton sounds like nothing so much as a time-travelling twenty-first-century geologist, somehow dropped three centuries into the past to do heroic battle with bad guys armed with dogmatic theology.

This can't be true. Even the word 'geology' wasn't commonly

used until *after* Hutton's death. I wanted to know how Hutton's great insights had come about, and how Hutton's ideas have been taken up and developed in the two centuries of geologising since his death, to become incorporated into our modern view of the world: I wanted to know the biography of the ideas, as well as of the man.

As I began to research more deeply, into academic works on Hutton and the (scarce) source material, at last I began to get a sense of the real Hutton. He was, of course, a man with a complicated personal life. He was deeply embedded, as we all are, in the culture and the great events of his time – the eighteenth century was an alien environment compared to our modern world. He was most resolutely *not* a modern scientist, in the sense we understand it now. But I saw that, remarkably, much of what shaped Hutton still attracts our thinking today.

This book, then, is the story of how a farmer's son from Scotland learned to peer into the deepest abysses of time. It is a drama of personality, landscape and ideas, of an intellectual revolution that shaped our world – and of a man whose vision, rooted in antiquity yet tinged with modern philosophies, was not only ahead of his own time but speaks to our new century.

At Siccar Point, Hutton peered into his friends' faces, earnestly seeking understanding. Gradually Hall and Playfair learned to see the deep history the rocks' formation implied. 'On us who saw these phenomena for the first time,' Playfair would write, 'the impression made will not easily be forgotten ... The mind seemed to grow giddy looking so far into the abyss of time.'

But Hutton, the man who saw eternity, had less than a decade left to live – and soon his most formidable opponent, Irish chemist Richard Kirwan, would launch a vitriolic attack against the vision that had shaped Hutton's life.

Stephen Baxter, June 2002

ONE

Deposition

I

'They make a desert, and they call it peace'

James Hutton was born in Edinburgh on 3 June 1726. He came into the world nine months before the death of Isaac Newton. He had three sisters, Isabella, Jean and Sarah, and an older brother who died when Hutton was very young. His father William Hutton was a merchant, prominent enough in Edinburgh circles to hold the office of City Treasurer for some years. John Playfair (Hutton's companion at Siccar Point, who would one day write Hutton's first biography) described William as 'a man highly respected for his good sense and integrity'. William's wife, Sarah, was a merchant's daughter.

Scotland's capital at this time was a city of some thirty thousand souls. It had yet to see the grand Georgian development of the second half of the eighteenth century, and it retained much of the character of medieval times. The Royal Mile – the avenue that still stretches from the castle down the rocky shadow of the glaciers – was universally admired as a beautiful fairway. The line of roofs that stretched from the castle to Holyrood Palace was punctuated by the steeple of the cathedral of St Giles, Parliament House, church spires, and the walls of a prison and university. But the unplanned clutter around the Mile was less revered: tenement blocks rose nine or ten storeys from the streets below, all in a cloud of smoke. The English chaplain of a Scottish regiment would compare the city to 'an ivory comb, whose teeth on both sides are very foul, though the space between them is clean and sightly'.

In the home of a middle-class merchant like William Hutton there would have been a dining room and a drawing room, cupboards of carved wood, panelled walls, upholstery of silk and scarlet leather, carpets and rugs, and stands of books. The Huttons were a conventionally Christian family, pious in their domestic way; they attended the kirk, and the children were encouraged to pray daily.

The family's deeper roots were probably southern. Hutton is a Yorkshire name, coming from an Old English word meaning, appropriately enough, 'settlement on a bluff'. But the Huttons spoke Scots. More than a dialect, Scots shared its origins with English in old Anglo-Saxon, but the two languages had diverged widely: Scots had been leavened with words from Scandinavian, French and Gaelic. The language had produced a rich literary heritage in the Middle Ages, and in Edinburgh, Glasgow and Aberdeen it had become the language of the Kirk, law and commerce.

However, an Act of the Parliaments passed only nineteen years before Hutton was born had united Scotland with England. Suddenly English was the language of letters and polite society. Learning English was as difficult as learning any new language: you had to remember to say 'old' instead of 'auld', 'a lot' instead of 'a muckle'. Scots was made to seem second-class.

The Scottish people did not bow down to English cultural imperialism. Scots continued to be spoken in the courts and the pulpits, and Robert Burns would be effectively bilingual, capable of composing verse in Scots as well as in English. Hutton himself would remain a steadfast Scots speaker throughout his life.

Scotland was proud, then – but in 1726 it was a backward nation. The highways were so poor that men and women alike would ride horses rather than suffer the discomfort of a coach. The fine natural harbours were underdeveloped, and in the economic shadow of England there was scarcely any shipping anyhow. The towns, including Edinburgh, still stank of poor sanitation and overcrowding, and were incubators of disease. There was poverty and hunger: the population had long outgrown the ability of the primitive farming methods of the time to feed it.

And it is remarkable to recall that even at this time, not very far to the north of Hutton's home in Edinburgh, a very different way of life persisted, and a language was spoken that was much older than either English or Scots. Like all of us, Hutton had been born on the transient surface of deep history.

Scotland first emerged into the light of history nearly two millennia before Hutton was born.

In the first century AD the Romans fought a great battle in the north-east of the country, against people described by the historian Tacitus as tall and red-haired. These Picts – after the Latin *picti*, for 'painted one', perhaps a soldier's nickname – were the product of some three millennia of previous waves of immigrants: flint-chippers from Ireland, England and Norway, Neolithic farmers from the Mediterranean, iron-users from across the North Sea. The bronze-working 'Beaker Folk' left a legacy of spectacular stone circles. The first named inhabitant of Scotland was Calgacus, the Swordsman, slaughtered in Tacitus' battle, and his angry defiance of the mighty Romans would characterise the Scottish history to follow: 'They make a desert, and they call it peace!' In the Picts' unmapped wilderness of mountains, bogs and forests, the Romans lost legions.

By the fifth century Rome was dying. In the turmoil that followed, more invaders came to Scotland, and there were endless wars between Picts, Teutonic-speaking Angles from across the North Sea, the Britons under their resolute kings (one of whom may have been Arthur) – and a wandering people from Ireland: they called themselves *Gael*, but the Romans had called them *Scoti*, 'bandits'. By the ninth century the Scots were overwhelming the Picts, who would be largely erased from history. The Scots established their seat of power at Scone, where kings were crowned over a slab of stone said to have been brought from the Irish homeland (though the geologists, prosaically, say it is local sandstone).

In their harsh land of granite and moor, the lives of the people were hard. To the perils of drought, flood and Norse raids was added the bloody froth of the endless battles of succession among the various kings and pretenders. One of them was Macbeth of Moray – in real life, evidently, a good ruler who may have made a pilgrimage to Rome.

In 1072 the Normans, fresh from their defeat of the English, mounted the most decisive invasion Scotland had seen since the retreat of the legions. Scotland itself became a Norman feudal kingdom, a great hierarchy of vassals and possession with everything ultimately owned by the king, who in turn was the vassal of God. The old Celtic tribal society nevertheless survived in the Highlands, mutating slowly into the clan system. The clan was seen as a kind of extended family, its people the 'children' of the chief, the head of the family. The Highlanders remained a warrior people, and among the clans there would be a long history of feuds, battle and treachery; there was little to unite them but their contempt of the Lowlanders, the *Gall*. It was a system that would, remarkably, persist until the eighteenth century, the age of Rob Roy, the Bonnie Prince, and James Hutton himself.

The economy slowly became mercantile and agricultural, but much of the country remained wild. A great forest blanketed the Stirling plain as far as West Lothian, and wild animals far outnumbered people – there were wolves, boar, herds of wild cattle and deer. The Scottish monarchy remained weak, plagued by battles over succession, and there were inconclusive wars with the English. The great events of Europe's Renaissance touched Scotland comparatively little: in an age of more weak kings, more inconclusive conflicts with the English, fratricidal bloodshed among the clans and banditry among the barons, money was spent on weapons and walls rather than patronising the arts.

In the sixteenth century Henry VIII's reign in England brought a breach with the Church of Rome. John Knox, a colourful and ferocious figure whose soul had been hardened by time spent at the oars of a French penal galley, provided Scotland's own Reformation with a focus. The Kirk of Scotland cut its ties to Rome, proclaiming that its General Assembly was above Parliament in authority. Even the young queen, Mary Queen of Scots, had to endure repeated confrontations with Knox over limits to her power.

For the ordinary people the Kirk's dogmatic severity, enforced by punishment and humiliation, was unwelcome. Old rituals, like summer plays in honour of Robin Hood, were condemned. Even dancing and drama were viewed with suspicion. Still, to visitors the people remained hospitable, with a capacity for drink in excess of that of any English, despite the Kirk's strictures. And Knox bequeathed a unique vision of political power. He believed that power was ordained by God – but it was vested in the *people*, not in the person of a monarch. The Kirk was hardly a proponent of modern democracy, but at the time its argument for a power vested in the common folk had no counterpart anywhere else in Europe. James Hutton would benefit from the Kirk's deep commitment to education.

Hutton's father died when James was three years old.

William's legacy seems to have left his family well provided for – he would even leave Hutton property, in the shape of two small farms in Berwickshire. After William's death his widow Sarah, with admirable strength of character, coped well; Playfair said she 'appears to have been well qualified for this double portion of parental duty'. Sarah resolved to bestow on Hutton a 'liberal education'. So Hutton was sent to the High School of Edinburgh.

Founded during the reign of Mary Queen of Scots, the school had been built from the rubble of a Blackfriars monastery, itself founded in 1230, and destroyed in 1558 by a Reformation mob. The monasteries had long served as seats of learning, and in 1566 it was resolved to build the High School on the site. The building was completed in 1578. By Hutton's day it was old and crowded. (It would be demolished in 1774 and replaced by a new building, which today houses the archaeology department of the University of Edinburgh.)

For its time, the High School was a good school. As early as 1560 John Knox had called for a national system of education, and the first parliamentary act to that effect was passed eighty years later. By Hutton's time nearly every parish had a school of some sort. In some places the education was no doubt rudimentary, but it was there, and it was free, at least in theory. By the time of Hutton's birth the literacy rate among Scottish males was more than half, and by 1750 virtually every town had a library.

The Kirk's motive for this educational drive was ideological: only literate children could read Holy Scripture. But a literate reader could not be confined to the Bible. The brilliant Scottish Enlightenment of the second half of the century would be nourished by the minds of an educated populace; David Hume and Adam Smith wrote not just for an intellectual elite but a genuine reading public.

What kind of child was Hutton? The biographical sources are very thin. We can infer from his later career that he was sociable and clever, and that he possessed a lively but restless and undisciplined mind. He probably had a distaste for protocol and authority. Much later, in 1774, he would write to his friend James Watt, developer of steam-engine technology: 'Your friends are trying to do something for you ... Every application for public employment ... requires nothing but a passage thro' the proper channels ... the honestest endeavour must to succeed put on the face of roguery ... come & lick some great man's arse and be damn'd to you.' At Edinburgh High School, the schoolboy Hutton must have been a handful for his masters.

And it would be delicious to know how old Hutton was when he first questioned the belief, held by most educated people, that the Earth was a mere six thousand years old.

2

'The first day of the creation is deduced'

James Ussher was an Irish bishop who had been chaplain to an executed king. He died in 1656 in England, far from home and all but bankrupt. But despite his troubles he completed his life's work, a mighty history of the Earth itself, called *The Annals of the World*.

Nobody today reads the *Annals'* densely argued Latin, spread across two thousand pages. But many people still know the *Annals* through one clear and unambiguous date and time: 22 October 4004, BC, a Saturday, about six in the evening. For this – so Ussher had calculated – was when God had created the universe, and history began. On that fateful Saturday afternoon, there was no four o'clock or five o'clock: it wasn't just that planet Earth didn't yet exist at those times, but those mundane instants themselves, hours now used for soccer-playing or tea-making or car-washing, *never occurred at all*.

For its precision and for the weight of scholarship that lay behind it, Ussher's date became imprinted in the mass consciousness for two centuries or more. In the early nineteenth century even Charles Darwin would graduate from Cambridge University believing that the world was six thousand years old, give or take.

But Ussher's intense and obsessive project seems very odd to the eyes of a modern scientist. After all he had deduced the age of the Earth without looking at a scrap of physical evidence – not a single rock.

James Ussher had been born in Dublin in 1581 into a Protestant family. Quiet and bookish, he was immersed in religion from birth. His uncle was the Archdeacon of Dublin, and Ussher learned to read the Bible with the aid of two blind aunts who had memorised much of its contents.

The times were turbulent. Ireland had long been a complex

web of Anglo-Norman fiefdoms and Irish kingdoms. During Ussher's lifetime Elizabeth I passed the Acts of Supremacy and Uniformity, imposing the Anglican Church settlement on Ireland. Her reward was three serious rebellions. The last rising, supported by the Pope and by Philip III of Spain, erupted in Ulster during Ussher's teenage years. Its suppression was a bloody business: English chroniclers spoke of the starving Irish creeping back from the forests where they had been driven, their mouths stained green by grass and nettles. From now on Ireland would be run almost exclusively by Englishmen, and Dublin was like a garrison city.

Against this background, young Ussher progressed with his studies. He was taught Latin at a school in Dublin run by two Scotsmen as a cover for their work as spies for the Scottish king James VI. By the age of ten Ussher had already become deeply pious, and soon after that fascinated by history – and by fifteen he had already made his first attempt to map out a chronology of the Bible.

At the age of twenty-one Ussher was ordained a priest. He would face a lifelong struggle against the Catholicism of the mass of the Irish people. Gradually Ussher came to see that he could use his Biblical scholarship, on the issue of the dates and wider matters, to support his Protestant faith. His intellectual reputation developed quickly. By twenty-four he was made Archbishop of Armagh; later he became Primate of All Ireland. Cushioned by wealth and position, Ussher settled down at the bishop's palace in Drogheda to work on his mighty history of the world.

The notion that the world might have a beginning at all is actually a legacy of Christianity. Most ancient civilisations had viewed the universe as eternal. Time was cyclical, with events repeating over and again – like the beating of a heart, the waxing and waning of the Moon, the cycling of the seasons. The Babylonians developed a cosmic model based on periodicities of the planets in which each Great Year lasts 424,000 years; in the 'summer', when all the planets congregate in the constellation Cancer, there is a great fire, and the 'winter', marked by a gathering in Capricorn, is greeted by a great flood.

Some thinkers had developed this notion to its extreme.

Perhaps, the Greek Stoics had argued, events repeat *exactly* from one cycle to the next. In a universe governed by *palingenesia* you have read these words an uncounted number of times before and are doomed to read them again, over and over, in future cycles. Aristotle was troubled by this notion. There would be problems with causality, he pointed out, if he found himself living as much before the fall of Troy as after it.

In cyclical universes it was impossible even to frame the question of an age of the Earth – for it hadn't had a beginning; there were only the mighty cycles, Great Years receding into past and future, beyond the reach of imagination.

There was one culture which eschewed the notion of a cycling eternity. In the Judaic tradition the history of the world was a narrative: a simple story, with a beginning, middle and end, spanning time from God's creation of the world on the first day all the way to the end of things. With the emergence of Christianity this story was elaborated further, with detailed revelations of what would come at the end of time, and with Christ's biography as a unique pivot.

St Augustine completed this great time-mapping project, arguing powerfully against the notions of cycling time. If life was doomed to follow patterns set in an earlier age, there would be no motivation to follow the teachings of Christ: what would be the point of trying to lead a better life, if every action you took was fixed before you were even born? And besides, cycles in time would violate one of Christianity's key precepts, that the Incarnation of Christ was a unique event. 'God forbid that we should believe in [the Eternal Return],' Augustine wrote. 'For Christ died once for our sins, and rising again, dies no more.' There were no previous cycles, then. Time was created with the world – which itself was born just before humanity – and would end with it. Augustine's faith surely consoled him in a difficult age. He had been born at a time when the Pax Romana seemed inviolable; he died with his city, Hippo in Africa, under siege by the Vandals.

Augustine's pronouncements on Earth's history froze ideas of time in western human minds for thirteen hundred years. The idea of cycles in time, and of a deep ancient history preceding humanity, faded from view, save as a half-remembered pagan notion. Earth's youth became powerfully lodged as essential to the faith, a doctrine it would be heretical to deny.

And remarkably, exactly *how* old Earth was could be worked out from a careful reading of scripture. The Book of Genesis contained a genealogy of Adam and his descendants through twenty-one generations, and the Bible's family histories continued through to dates anchored in recorded history, like the destruction of Jerusalem's temple by the Persians. So in principle, as the young Ussher understood, to get a date of divine reliability for the beginning of time, all you had to do was to search through the Bible for the relevant generations and add them all up.

Ussher's early adolescent experiments quickly taught him that in practice this was tricky. Scholars had actually been trying to construct Biblical chronologies since Roman times. While these computations all gave results of the same order of magnitude – a few thousand years – they differed unfortunately widely, ranging from less than six thousand years to nearly nine thousand. To get to the 'true' answer was going to require some careful scholarship.

As a firm basis for his work, Ussher sought out the most authoritative version of the Bible. All editions of the Bible had been copied and translated from earlier versions, and Ussher understood that the older the copy, the less likely it was to be corrupted by translation and other errors, or even deliberate falsifications. He employed an agent to seek out rare manuscripts in the Middle East, and went so far as to learn the ancient languages of Samaritan and Chaldean when scraps of Biblical scripture turned up in those languages. He already knew Hebrew.

Meanwhile, Ussher tried to link the Bible's chronology to known dates in history. He was greatly helped by the ancient chroniclers' habit of referring to astronomical events. The motions of the stars, planets, Moon and sun are perfect timekeepers: events like solar and lunar eclipses are comparatively rare but they can be forecast – or fixed in time retrospectively, using the same techniques – with great precision. Thus Ussher used an eclipse to fix the birth date of Christ.

Ussher laboured over his studies for decades, even when he became chaplain to the king of England, on the accession of

Charles I. But as he approached old age Ussher's immersion in the past was to be interrupted by a wave of violence in the present.

As the storm clouds of the English Civil War gathered, the autocratic King Charles sent new English colonists into Ireland, and set up an Irish Army. His most bitter enemies, the Puritans in Parliament, declared that the purpose of these moves was the invasion of England and imposition of a royal tyranny, and Parliament ordered that the administration of Ireland be passed to Puritan lords justices.

For the Irish Catholics there couldn't have been a worse set of rulers than these grandees, and a general rising became inevitable. Ussher, as a Protestant landowner, was a target for the fury. His chaplain at Drogheda was threatened with burning, with Ussher's precious books to be used as faggots beneath his feet. Ussher himself had the good fortune to be in England at the time of the Catholic eruption, and thus escaped seeing his country houses plundered, his herds and flocks killed or dispersed. However, Ussher could never return to Ireland – and the turmoil left him broke, his income stream gone.

After Cromwell's military victory in the Civil War a union of the kingdoms of England, Scotland and Ireland was imposed. Ireland, a conquered country, was parcelled out among the soldiers and creditors of the Commonwealth. Meanwhile the king was brought to London. Ussher, watching from the roof of his protector's house in Charing Cross, fainted dead away when Charles laid his neck on the executioner's block.

Even in the midst of this turmoil Ussher had doggedly continued his work. At last he found the key historical link in his scrutiny of the Bible. It was a bland-looking reference to the death of Nebuchadnezzar, king of Babylon – but Ussher could use this to link the Bible's chronology to Greek history, and then via Roman history to the modern calendar. Ussher was thus able finally to move on to an authoritative date for the beginning of the world: 4004 BC, exactly four thousand years before the birth of Christ himself.

As for the date and time, it was accepted that God would have created the world at a solstice, as such a moment of astronomical symmetry seemed appropriate as the instant of creation. Since the Garden of Eden was well stocked with fruit when Adam and Eve first awoke, Ussher decided this must have been the autumn equinox in October – and surely a Sunday would be the first full day. Hence his selection of a Saturday afternoon for the instant of creation itself, for then it would follow that, according to the Genesis verse, 'the evening and the morning were the first day'.

Thus, Ussher wrote triumphantly, 'the first day of the creation [is] ... to be deduced'.

I can't quite imagine having in my world a text containing truths, not derived by any human process, but transmitted from a higher authority. Perhaps Ussher's Bible was like an encyclopaedia received by radio from some advanced extra-terrestrial civilisation, as in Carl Sagan's novel *Contact*. Certainly, for later generations, giving up such an authoritative source of information was going to be hard. As the Victorian naturalist Philip Gosse said (so his son reported), 'If the written Word is not absolutely authoritative, what do we know of God? What more than we can infer, that is, guess – as the thoughtful heathens guessed – Plato, Socrates, Cicero – from dim and mute surrounding phenomena': not a bad summary of the uncertain world view of modern science.

Whatever you think of Ussher's assumptions, though, you can't argue with the quality of his scholarship.

In 1656, aged seventy-five, James Ussher died. Though he had clearly been a Royalist, Cromwell recognised him as the fore-most Biblical scholar of his time, and honoured him with a state funeral and burial in Westminster Abbey – though the honour didn't extend to paying the bill. By the 1670s Ussher's dates were being printed in new editions of the Bible, and by 1701 the use of his chronology had been authorised by the Church of England. Soon the dates became as familiar a part of the Bible as the ancient texts themselves, and Ussher's chronology developed a theological weight. Ussher's was a cosmogony constructed on a human scale. Even if the idea of being trapped inside such a rapidly decaying universe was claustrophobic, it was a comforting, even a cosy thought, that behind us lay such a short morning.

Ussher's dates would no doubt have been included in the

Hutton family Bible, and his 'Mosaic chronology' (after Moses) would always form part of Hutton's thinking. He knew he would have to present evidence to argue successfully against the Biblical scholarship.

But Hutton was not the first to question Ussher. Even as Ussher was publishing his great work, doubts were raised. During the previous centuries Europeans had begun to travel the world. And they encountered cultures which had their own historical narratives, many of them contradicting the Biblical account. The Chinese, for example, mocked the story of Noah's Flood, which was supposed to have occurred around 2300 BC. Chinese written history stretched back centuries *before* this date, and made no mention of a disastrous global deluge. This news was received with great hostility in Europe, where it was imagined that the Chinese must have exaggerated their timescales as a matter of cultural prestige, but Jesuit missionaries in China believed the records of their hosts, and despaired.

By the time of Hutton's youth it was obvious there was something wrong. You didn't even have to look at the rocks to know that.

3

'A mind formed for different pursuits'

In November 1740, at the age of fourteen, Hutton entered the
University of Edinburgh, where he was to study 'humanities'.
Edinburgh was a packed, lively, exciting place to be.
Compared to London it was still a small city, and the folk
were crammed into the dank and dark wynds of the Old
Town. But it was full of energy. James Boswell described
running home after class, past 'advocates, writers, Scotch
Hunters, cloth-merchants, Presbyterian ministers, country
lairds, captains both by land and sea, porters, chairmen ...' In
the taverns there were regular performances by amateur
musicians on harpsichord, violin and especially the flute,
which was very popular. There was even dancing. The
aristocrats would mount 'assemblies' in their homes, of an
elegance to match anything to be seen in London or Paris, and
since the 1730s it had been possible even for ministers of the
Kirk to learn to dance without ostracism. In taverns and
hotels, the claret flowed by day and night. For the teenage
Hutton, revelling in his studies and his new social life, it must
have been a very heaven.

Hutton had once again benefited from his country's historical
legacy. The University of Edinburgh, founded in 1582 by James
VI, had been Scotland's first post-Reformation college. And as
the vast post-Union hangover suffered by the country as a
whole slowly dissipated, the old theological monopoly on the
curriculum was broken. Under a new liberal regime there were
professorships in medicine, law, mathematics and 'natural
philosophy' (the sciences). The Scottish universities were still
small, but drew students from across Protestant Europe, since
religious restrictions barred many students from the great
English colleges.

Meanwhile, university education in Scotland was remarkably cheap and accessible. The sons of farmers and

shopkeepers and builders, some, like Hutton, as young as thirteen or fourteen, would enter the universities alongside the sons of aristocrats and landowners: perhaps half the students at the University of Edinburgh came from middle-class backgrounds. To get in, all you needed was a knowledge of Latin, an ability to pay a fee of perhaps five pounds a year – and to be male, of course.

Hutton, though, had a wayward mind. One day, bored by a lecture on logic, he was distracted by a chance remark from the professor on how acids may be used and combined to dissolve metals. The professor was trying to illustrate some wider philosophical point, soon forgotten. But for Hutton, this sparked an attraction to chemistry's mysteries.

Perhaps, as Playfair would speculate, chemistry's appeal to Hutton was that 'Nature, while she keeps the astronomer and the mechanician at a great distance, seems to admit [the chemist] to ... a more intimate acquaintance with her secrets.' Certainly Hutton's hands-on exploration of chemistry would 'decide the whole course and complexion of his future life' – and would eventually prove a source of significant income.

Hutton began to search for material on his new passion, but pursuing academic interests wasn't so easy in the mid-eighteenth century. There was no TV, no Internet: there would be no *Encyclopaedia Britannica* until 1768. There wasn't even a separate chemistry curriculum at the university.

The only way Hutton found to follow up his new interest was through John Harris's *Lexicon Technicum*. Published in 1704, this 'Universal English Dictionary of Arts and Sciences' was actually the first encyclopaedia approaching a modern form to be published in English. The *Lexicon* was important in that it emphasised scientific and technical subjects, and contained clear engraved plates, practical text and bibliographies. It was incomplete by modern standards, however: it had no cross-referencing, for example.

Meanwhile, Hutton was lucky enough to study mathematics under Colin Maclaurin. Maclaurin, aged just forty-two when Hutton entered the university, was one of Britain's foremost mathematicians. Born in Argyllshire, he was a child prodigy who had entered the University of Glasgow at the age of eleven, was a full professor by nineteen, and a member of the Royal Society of London by twenty-one. Maclaurin's name is still known today to millions of mathematics students through the 'Maclaurin series', a way of analysing mathematical functions.

Hutton wasn't so impressed at first. He professed to admire Maclaurin's lectures, but he 'cultivated the mathematical sciences less than any other'. However, the teachings of Maclaurin would shape his thinking profoundly – for Maclaurin was Hutton's link to Newton.

Sir Isaac Newton, physicist and mathematician, had been the towering figure of the previous century's scientific revolution. Maclaurin had become acquainted with Newton through the Royal Society, and it was Newton who had recommended Maclaurin for his professorship in Edinburgh – indeed, Newton offered to pay a contribution towards Maclaurin's salary. Maclaurin would go on to extend Newton's work in calculus, geometry and gravitation, and through Maclaurin, Edinburgh would become a centre for the teaching of Newton's work, and his broader philosophy.

Newton had followed in the footsteps of René Descartes. A shy Frenchman who had been forced to conceal his radical thinking from the savagery of the Inquisition, Descartes had tried to develop a new kind of philosophy based on reason. With such methods, 'there is nothing so remote', Descartes said, 'that it cannot be reached' – not even the birth of Earth, on which Descartes speculated. Descartes's work set out the first modern model of the universe as a place governed by a few simple laws. Newton would build on this, replacing Descartes's undefined principles of nature with precise and mathematical rules of motion, force and gravity.

But the new 'mechanical philosophies' didn't just give students like Hutton new tools to study Nature. They gave them a new way to think about God.

'Newton infers, from the structure of the visible world, that it is governed by the One almighty, and All-wise Being,' Maclaurin would write. 'The simplicity of the laws that prevail in the world, the excellent disposition of things, in order to obtain the best ends, and the beauty which adorns the work of nature, far superior to any thing in art, suggest his consummate Wisdom.' Maclaurin's students were taught that the world around us is not the work of the Devil or a ruin of some more perfect past time, as conventional readings of scripture would have it, but an orderly and bountiful place that reflects the generosity of God and the regularity and symmetry of His natural laws.

This 'Deist' thinking, a major break from the orthodox religious traditions of the time, had originated in seventeenth-century France, and flourished in England in the first half of the eighteenth century. The Deists accepted the Bible's moral authority but rejected its literal interpretation as a true history of the world. God wasn't banished, but His role was restricted to setting the universal laws. He emphatically did *not* tinker day to day in the working of the world, through miracles and Floods. Newton's vision of a world governed by simple laws led to its natural incorporation into the Deist vision, even though this was to some extent a misinterpretation of what Newton himself believed. Later Deism became well rooted in revolutionary America, counting Benjamin Franklin and the first three Presidents among its adherents.

This kind of thinking and its uplifting message – that the world is an orderly place, governed by comprehensible and unchanging forces in balance, and reflecting a divine design – had a profound impact on the young Hutton, and its influence would show in his later scientific work. In language clearly echoing Maclaurin's, he would write four decades later of 'the globe of this Earth as a machine, constructed upon chemical as well as mechanical principles, by which its different parts are all adapted, in form, in quality, and in quantity, to a certain end; an end attained with certainty or success; and an end from which we may perceive wisdom, in contemplating the means employed.' A seed of religious doubt seems also to have been planted, which would come to full flower later in his life.

The brilliant Maclaurin must have seemed a Moses to his students, a personal acquaintance of the God-like Newton already more than a decade dead.

Hutton prospered at university, his mind flowering, intellectual avenues opening. 'His taste and capacity for instruction were sufficiently conspicuous during his course of academical study,' said Playfair. But after three years of study, Hutton had to leave.

William Hutton had left his family comfortable but not wealthy. His son, as the only male in a family of five, was the most likely to gain decently paid employment, and, just seventeen years old, he had to shoulder his responsibilities. Hutton's college enthusiasms did not offer a way forward as a possible career. The Industrial Revolution had scarcely begun, and it would only be in later decades that chemistry came of age as a practical subject, largely spurred by the need to replace natural bleaching techniques based on sunlight, rain, sour milk and urine. Science itself was still more a hobby of independently wealthy men than a career with a well-defined path. So, though going into business 'was by no means congenial to his mind', Hutton gave in to pressure from his family and friends and did just that.

He was placed as an apprentice to a 'writer to the Signet' (what the English would have called a solicitor – the job was named after the royal signet used to authorise legal documents). As Playfair said, 'Under the subjection of the routine of a laborious employment, [Hutton] was now about to check the ardour and repress the originality of a mind formed for different pursuits.' Hutton soon found himself buried in dry-as-dust detail.

Even as a legal clerk, though, young Hutton might have found some intellectual stimulation. The Scottish legal tradition had been jealously guarded through the negotiation of the Act of Union. Though the two systems had sprung from the same root in the twelfth and thirteenth centuries, English law had become inward-looking, relying on a body of precedent, the mass of previous decisions handed down over the generations. But the Scottish lawyers looked to the Continent, where the ancient principles of Roman law had been rediscovered and revived: Scottish judges were encouraged to rely not just on precedent but on reason, based on principles of fairness and justice. Thus, even in the law, Scotland was becoming a land of reason, not dogma.

But the law, Scottish or otherwise, wasn't for Hutton. Once again his mind refused to stay disciplined. His employers found him amusing himself and his pals with chemistry experiments during work time, when he should have been studying legal forms or copying papers. Hutton was fired – or as Playfair tactfully tells us: 'With much good sense and kindness, therefore, [his master] advised [Hutton] to think of some employment better suited to his turn of mind.'

Young Hutton now settled on a new career path: in medicine.

At first glance this was a good choice. Medicine was an honourable profession, it was (and is) perpetually secure, and, as Playfair noted, it was 'the most nearly allied to chemistry'. Hutton began his studies under a doctor called George Young, and he started to attend lectures at the university again, just a year after leaving it the first time.

Back at the university Hutton quickly became fast friends with a fellow medical student known as John Clerk of Eldin. The Clerks were distinguished for their intellectual ability and public spirit. Sir John Clerk, Eldin's father, had actually been one of the commissioners who negotiated the Union with England. As a younger man, though, Sir John had flirted with more seditious tendencies; during his Grand Tour of Europe in the 1690s he had rubbed shoulders with cardinals and contessas and nearly became a Catholic. On his return to Scotland he had realised where his interests lay, and from the beginning of the eighteenth century to his death, he never wavered in his belief in the rightness of the Union for England and Scotland alike. Eldin's brother George, who would become known as George Clerk-Maxwell, was nicknamed 'the late king's Godson' since he had been born in the year of the 'Fifteen – the abortive Jacobite uprising of 1715 – and had been loyally named for George I.

The Clerk family would remain close to Hutton throughout his life – and they quickly proved an important influence, for they were wealthy landowners who for generations had mined coal from their extensive Midlothian tracts. As mine owners, they had a strong practical interest in the structure and riches of the Earth. This period, then, was not just the start of Hutton's medical career, but perhaps also marked the birth of his interest in geology. He was still only eighteen.

What was the student Hutton like? His only contemporary biographer, Playfair, would know him only in old age, and of this period Playfair tells us only that Hutton 'pursued with great ardour the studies of chemistry and anatomy'. But we can be sure that young Hutton was a lively, stimulating and gregarious companion, who made friends easily: 'His conversation was extremely animated and forcible, and, whether serious or gay, full of ingenious and original observation ... A brighter tint of gaiety and cheerfulness spread itself over every countenance when [Hutton] entered the room.' Playfair, moreover, had an interest in sanitising the image of Hutton. We know from Hutton's letters that he was in later life a lusty man, fond of drink, eating well and of other physical pleasures – but Playfair is silent on all this. He also liked bawdy jokes! In a letter to the engineer James Watt written in 1774, Hutton said a local entrepreneur had been consulting him about a patent for a new improved bed: 'I'm thinking of adding to it a machine which shall be called the muscular motion whereby all the several parts shall be performed of erection, intrusion, reciprocation and injection ...'

At the age of eighteen, it was surely not just chemistry and anatomy that attracted Hutton's ardour.

While Hutton's young life enjoyed a new beginning, however, his country was about to undergo great anguish.

On 23 July 1745 the Young Pretender, Prince Charles, returned from exile overseas to land in the western Highlands. He brought with him seven men, no money and no weapons. Even though by now only the oldest Scots could actually remember a time when a Stuart king had occupied the throne, their romantic legend burned fiercely, and within weeks the Prince had gathered an army of Highlanders.

Whether you called what followed a 'rising' or a 'rebellion' depended largely on whether you were Jacobite or Hanoverian. The military authorities were surprisingly unprepared. Even the Highland forts and barracks – completed at considerable expense since the last Jacobite rising – were so undermanned that their garrisons could do nothing but watch as the Prince's ragtag army marched past on their way south.

Charles enjoyed some rest and recreation at the ancient

Stuart palace at Linlithgow, where he made the fountain run with red wine. Then the Highlanders marched, unopposed, on Edinburgh.

4

'Every man was a soldier'

The roots of the 'Forty-Five reached back to the aftermath of Union.

After a great deal of agonising the Act of Union had been passed in 1707. Even the negotiations had caused widespread civil unrest; for a time Edinburgh was placed under martial law. But the more realistic of Scotland's rulers understood that the country had effectively long been run from England anyhow, and that they may as well seek the benefits to trade that a formal union would bring. Edinburgh had become a capital without a king, court or parliament – but the preservation of the country's law, church and universities would ensure that Hutton's Edinburgh would be well stocked with lawyers, scholars and clergymen, but mercifully few professional politicians.

The immediate aftermath of Union was deeply unhappy, however. Suddenly customs and excise duties were imposed from London, at much higher rates. For a time smuggling became almost a patriotic act, and customs officials – especially if they were English – could face personal danger as they went about their work. What made it all worse was the attitude of the English. When Scottish members of the London parliament asked for an equality of taxation treatment in Scottish linen and English wool, the Lord Treasurer replied, 'Have we not bought the Scots, and a right to tax them?'

Discontent in Scotland found a focus in the continuing existence 'over the water' in France of Stuart pretenders to the throne: Charles I's son James Edward, the 'Old Pretender' (the word 'Jacobite' comes from the Latin Jacobus, for James), and James' son Charles Edward, the 'Young Pretender'. The allure of the Stuarts was magnetic: they represented, at least in hindsight, a simpler, more stable time. The Stuarts even had supporters among the intellectuals, including Samuel Johnson and Alexander Pope. There were abortive Jacobite risings in

1715 and 1719. But the 'Forty-Five was a different matter.

The War of Jenkins's Ear (arising from a clash between the Spanish authorities and one Captain Jenkins, an English privateer off the coast of Cuba) had enmeshed Britain in a European war with Spain's allies, including France. The British, desperate for troops, stripped garrisons in northern England and Scotland. Realising this, the French made plans to land a force in Scotland, thereby opening a second front. Prince Charles, twenty-four years old, was charming, handsome, personable, and the French saw him as an ideal figurehead for this adventure. When a storm scattered the French invasion fleet – a 'Protestant wind', as some called it – the French grew cool, Charles frustrated. Charles concocted a new plan: to land in Scotland with only a few supporters and to raise an army himself. Even loyal followers called this a 'mad enterprise', but Charles was driven by a reckless sense of personal destiny.

In the Highlands, Charles's arrival was a spark to dry tinder.

In the glens, the clan system, that strange, anachronistic echo of pre-feudal times, still functioned. Modern standards of life dissipated as you travelled north: the clan chief's rule remained absolute, even over life and death. The Highlanders lived in one-room houses of mud and stone, called bothies. From a distance you would think a typical village was just a huddle of mounds of dirt: it was a shock to realise that these heaps housed human beings – even though by now their chiefs wore lace and drank claret, and sent their sons to the Lowland universities.

By 1745, though, the Highlanders were on the brink of starvation. Six hundred thousand people lived on glacier-scraped soil which, given the farming techniques of the time, could support perhaps only half that number. The theft of cattle from neighbouring clans had become an industry, even a necessity for survival. But in the Highlands, as Johnson said, 'every man was a soldier', and now a Stuart was calling them to arms. The burning crosses, the ancient summons of the clans, blazed from mountain to mountain.

John Maclean, a typical volunteer, was a Maclean of Kingairloch, a clan whose main territories were on the island of Mull. An officer of the Black Watch in his thirties, he decided to switch sides to the Prince. He travelled across the

Sound of Mull and met the Prince at Kinloch-moidart where, he recorded in his journal, 'I had the honour and Satisfaction to Get a kiss of his royal Highness his hand.' Charles made Maclean a captain; Maclean would follow the Prince all the way to Culloden.

In Edinburgh, and in Glasgow and Aberdeen, there was no desire to see Charles succeed. In the cities the Union had gradually brought affluence and prosperity, just as its wiser designers had hoped. The new urban middle classes were marching to a better future, but the return of the Stuarts would mean a reversion to an older Scotland.

But now a Stuart army approached Edinburgh itself.

The military situation was grave. Edinburgh Castle was a fortress housing a royal garrison, and was the base of the British government's Scottish Command. But a foreign war had bled the garrison down to three thousand troops, most of them inexperienced or of doubtful loyalty. Then, as Charles's army approached, the English commander in Edinburgh withdrew altogether.

Suddenly, incredibly, Edinburgh was left defenceless. The citizens were thrown into a state of panic, wonder and excitement, 'all being in the greatest flurry and confusion', as one observer said. Many of the city's establishment fled to Berwick – where, bizarrely, they indulged in betting games on the progress of the rebels. The town council emphatically did not rise to the occasion. They showed no desire, and formulated no plan, to oppose the approaching army.

Into the breach stepped two private citizens: a merchant called George Drummond – and Colin Maclaurin, the Gaelic-speaking disciple of Newton and professor of mathematics to Hutton. Drummond and Maclaurin called for volunteers to assist the undermanned garrison. Many students came forward. The 'College Company of Volunteers' were academics and clerics, the pillars of the Scottish Enlightenment in the decades to come. And now they were putting their lives on the line to save the city from the advance of the 'wyld heighland men'. We don't know if young Hutton was a volunteer. But, caught up in the swirl of these exciting events, Hutton must have thrilled to see his old maths professor assuming command in such a way.

The students were joined by elderly men wearing the great periwigs of their youth, and country folk carrying fowling pieces and water canteens as if ready for a day on the hills. It was a motley bunch, and their preparation was amateurish. They were stood up and stood down in turns in what they called a 'Grand Old Duke of York' game. Still, they dug out what armaments they could find, including cannon of various vintages that they set up on the city walls and, dressed in their great watch-coats, they began picket duty. Maclaurin designed improvements in the city's defences and vigorously supervised the work. He was assisted in this by Robert Adam, a future architect of note, then just seventeen. But the fortifications, in large part, were no more intimidating a barrier than a garden wall.

As time wore on, no reinforcements arrived from England. And on 15 September the city learned that the Jacobite army was only eight miles away.

The advancing Highlanders evoked genuine fear. They carried broadswords and scimitar-like 'Turks' for close-quarter fighting, and wooden shields called targes, studded and bossed with the interiors still bearing animal hair. They wore their tartans in complicated arrangements of kilted shirts, breeches and sporrans. Into their belts were hooked pistols, dirks and daggers. They had cavalry, including hussars, a type of unit new to British armies. With their furred caps, long swords and billows of plaid over their shoulders, the hussars would form a ferocious guard around the Prince as he travelled.

While the leading men of each clan were well armed, their peasant followers, coming from a background of extreme poverty, were less well equipped: some had nothing more than stakes pulled out of hedges. Even this, though, added to the menace. As Walter Scott would write, 'The grim, uncombed and wild appearance of these men ... created surprise in the Lowlands, but it also created terror.'

On that bright September day, the Edinburgh volunteers assembled and marched through the Old Town to meet the oncoming Highlanders, flags flying and drums beating – but most of the citizens barred their windows. Those who did watch the volunteers go greeted them with jeers, insults or even tears. Many looked, it was said, as if they were being taken to their execution. And here came the principal of the university to urge the 'flower of the youth of Edinburgh' not to waste their lives. Nerves weakened.

As the volunteers neared the West Port, their commander, Drummond, turned around to find his troops had disappeared, melting away into the wynds and taverns. The volunteers' resistance had ended before it had begun.

The next day, one of Charles's reconnoitring parties snuck into the city through a carelessly opened gate. The castle, with its garrison, remained secure, but the rest of the city fell without most of its inhabitants even knowing it, until they woke the next morning to find fierce-looking Highlanders manning the walls. Government officials hastily fled to the safety of Berwick. Maclaurin fled to England, to be sheltered by the Archbishop of York. Perhaps it was well that the city fell without a fight; it was largely spared any retaliatory destruction.

Charles would occupy the capital for five weeks. He established his court at Holyroodhouse, and his army made its camp in Holyrood Park, in the lee of Arthur's Seat. There was much softening of resolve as the Prince's troops were seduced by the city's taverns, and the palace buzzed with excitement as the Pretender gave assemblies, entertainments, and even a grand ball. Many of Edinburgh's ladies were said to have declared for Charles, captivated by his good looks and gallantry and the romantically heroic way in which he had thrown himself on the mercies of his countrymen. It was a woman who first called him the 'Bonnie Prince'. But there were some uglier scenes. In one of the infirmaries wounded Highlanders were set upon by a mob of citizens who tore open raw wounds and twisted arms and legs that had been set after fractures.

Sir John Clerk, the father of Hutton's college friend, had served in the government militia during the 'Fifteen. Now sixty-nine years old, he had left his estate at Penicuik in the hands of his sons and made for Durham. When the rebels came to Penicuik they demanded hay and oats, under penalty of burning the house down. However, Sir John wrote, 'When the Highland parties came they were civilly used and so committed no disorders about the House except that they eated and drank all they cou'd find, and called for everything as they thought fit, for they lookt on them selves as the Masters of all the Country.'

Hutton, trying to go about his business in the city, would have had to get used to Highlander sentries on the streets. Some of them had brought their families and other camp-followers. You might see a Highlander having his hair deloused by his woman, accompanied by a screeching brood of children. The poorer troopers were often reduced to begging openly in the streets. For all concerned, it was a strange, atavistic time.

Charles's next military move was to set off south and try to foment rebellion in England. By 9 November the army had reached the boundary between the kingdoms. Charles's 'Highland savages' astonished the English as they marched through Manchester, Lancaster, Derby – through what, in fewer than fifty years, would become England's industrial heartland. At Derby, Captain John Maclean 'missed the Seeing of a Curious Silk Manufactory which ... had (as I was told) more than ninety thousand motions'. By December, Charles was just 130 miles from London.

But English support for the Jacobites, seen as essential to sustain the rising, was always fitful. And by now three armies were closing on Charles – including one commanded by the Duke of Cumberland, the King's brother and a veteran of European wars, in whose forces Hutton's friend George Clerk-Maxwell now served. To evade their enemy the rebels were subjected to forced marches of fifteen or twenty kilometres a day, in harsh wintry conditions. At Derby, Charles's advisers forced him to accept the inevitability of falling back.

The retreat was ugly, however. On the way south Charles's troops had behaved well, but now they lost their discipline, and their looting – and rumours of the slaughter of wounded English soldiers – left a legacy of rage. In Glasgow the merchants had their revenge for the disruption to their trade. They raised a regiment of militia which attached itself to the forces of government troops and diehard volunteers who now converged to retake Edinburgh. The capital was liberated on 6 January 1746.

On 16 April Charles, out of money and supplies, not even on speaking terms with his field commander, drew up his troops for their final stand at Culloden. It would be the last clash in

Europe between a modern army and a pre-modern force; the Highlanders' last charge broke on English numbers, discipline and technology.

In the Highlands, the aftermath was bitter. Cumberland was convinced that only radical and extreme action would finally root out the Highlanders' threat to the future of the Union. It began on the field of Culloden itself. In Maclean's journal it is recorded that, 'In this Battle the greatest barbarities was Committed that ever was heard to be done by Either Christians Turks or Pagan, I mean by our Enemies who gave no quarters Killed our men that was wounded in cold blood and continued so doing for three or four Days or any others they could Catch.' These words were actually written by John Maclean's kinsman Donald McLean, for John himself was killed during the battle.

And then the cleansing of the glens began, an action even mighty Rome had not had the resources to see through. At Greenock, a young James Watt saw his father's workshop searched by troops pursuing the Prince. Five years after Charles Stuart had fled to France, kilted fugitives were still being hunted by armed patrols. The son and brother of kings, Cumberland would earn himself the title of the Butcher.

Little of this remote tragedy touched the cities. There, Culloden was portrayed as the defeat of an Antichrist. In Edinburgh, Drummond replaced the disgraced Lord Provost. The young volunteers were suddenly the heroes of the hour. Colin Maclaurin returned from York. Unfortunately the ordeal of his flight had damaged his health, and he would die in 1746, at the age of forty-eight. His premature death was a significant loss; in subsequent years Britain lost much of its influence in European mathematics.

Hutton's biographer Playfair, a gentleman scientist writing of these times sixty years later, would make no mention of the whole incident. To Hutton's generation the 'Forty-Five must have seemed embarrassing, a spasm of romantic and anachronistic nationalism which had claimed, among very many other casualties, a protégé of Newton – and had besides got in the way of business. After this great interruption Hutton got back to work at his studies, surely agreeing with Sir John Clerk, who would write, 'The success of [Culloden] gave universal joy, especially to friends of the Government, but there were even Jacobites who were at least content at what had happened, for peace and quietness now began to break in, whereas Anxiety and distress of various kinds had possessed the breasts of most people ever since the Rebellion broke out. All Trade and business in this Country were quite at a stand.'

5

'The Earth's blood is the veins of its waters'

Edinburgh folk were determined that their city would become distinguished in medicine, as in so many other areas of life.

John Monro, an army surgeon, was a conspicuous example: he deliberately programmed the education of his son Alexander to ensure that the boy would one day become professor of anatomy at the University of Edinburgh, and sent him to study at the University of Leiden in Holland, the acknowledged centre of European medicine. The father's ambitions were fulfilled; Edinburgh would become the leading academic centre for medicine in Britain, and Scottish doctors would become pioneers in surgery and obstetrics. James Lind, a naval surgeon from Edinburgh who would become a close friend of Hutton, would introduce fresh fruit to the British navy to prevent scurvy. Another Scotsman, James Pringle, made essential recommendations regarding the welfare of troops that would lead to the establishment of the Red Cross organisation in 1864 – and three generations of Monros would indeed teach anatomy at Edinburgh.

Despite these advances, much of the medicine studied by Hutton in the 1740s would have seemed primitive to us. The importance of public health and hygiene was only slowly becoming understood. The science of pathology had yet to be born, and vaccination (though long practised in the east) would not become well established for fifty more years. In some quarters the mentally ill were still held to be possessed by demons. In Edinburgh, controversy raged over the teaching of the writer John Brown: that there were fundamentally only two diseases, sthenic (strong) and asthenic (weak), and only two treatments were therefore ever required, stimulant and sedative – Brown's own preferred remedies were alcohol and opium.

No wonder Hutton developed a sour view of the profession.

He would write in 1771, 'The more medical knowledge we acquire, the more we know how little efficacious that art is.' Nonetheless, he persevered, continuing his studies through the period of the rebellion.

As Playfair would acknowledge, in Hutton's day the Edinburgh medical school was 'neither in reality, nor in the opinion of the world, so complete as it has since become [by 1805]. Some part of a physician's studies was still to be prosecuted on the Continent.' So after three years at Edinburgh, Hutton went abroad to complete his medical education. His first stop was Paris. He arrived there in 1747, aged just twenty-one.

It was quite an adventure. This was the Paris of more than four decades before the Revolution: the city of Louis XV, his mistresses and his scheming courtiers, a great and growing metropolis of more than six hundred thousand people. The Louvre, the Tuileries and the Champs-Elysées were all there to be marvelled at, and the eastern stretch of the Grands Boulevards was a fashionable promenade with many little theatres and cafés. Such splendour must have made Edinburgh seem small and provincial indeed. The political atmosphere seems to have been relaxed, even though Hutton's stay actually overlapped with a minor war between Britain and France called King George's War – a convoluted affair deriving from conflict over the Austrian succession and fuelled by colonial rivalries in North America.

For a lively, intelligent, gregarious, lusty young man like Hutton, far from his mother's watchful eye, Paris must have been marvellous.

After two years in Paris Hutton made his way to the Low Countries, where he would take his degree of Doctor of Medicine, at no less a centre than the University of Leiden itself.

Leiden was a city some thirty kilometres south-west of Amsterdam and eight kilometres inland from the North Sea. The old town was criss-crossed by a network of canals. The dominant textile industry was in decline, leading to an economic stagnation that would not end until the industrialisation of the late nineteenth century. For Hutton, Leiden probably wasn't as much fun as Paris – but its reputation in medicine remained unimpeachable.

Leiden's distinction derived largely from Hermann Boerhaave, who had served at the university as professor of botany, chemistry and medicine. Boerhaave was the founder of the modern system of teaching medical students at the patient's bedside; breaking away from medieval traditions, he encouraged his students to use their eyes and ears in their diagnoses. Boerhaave had been a great admirer of Isaac Newton and his orderly scientific thinking, and he believed that such principles should be applied to the study of the body and its workings. Boerhaave's teaching attracted students from all over Europe, and he had exerted an influence on the development of medicine in Vienna and Germany – and, through Monro, in Edinburgh.

Leiden, then, was a good place for Hutton to complete his medical training. And the academic thesis he produced to support his degree showed how, even by the age of twenty-three, much of the thinking that would characterise his later work was already maturing.

Hutton's thirty-four-page thesis, submitted on 12 September 1749, was written in French. Its title was *De Sanguine et Circulatione in Microcosmi* – 'On the Circulation of Blood in the Microcosm'.

In the seventeenth century, William Harvey's discovery of the reality of the circulation of blood had been the greatest medical triumph of the new post-Newtonian orderly scientific thinking. But Harvey's inspirational idea was not so much scientific as theological, for in the body's workings Harvey saw evidence of *purposeful design*. Such arguments seem to have struck Hutton deeply, and now he explored Harvey's thinking.

Design arguments had a deep tradition, reaching back to Aristotle. What, Aristotle had asked, is the *cause* of the existence of, say, your house? There is more than one cause. First there are the bricks and mortar of which it is constructed. These make up the *material* cause, without which the house obviously could not exist. But somebody must do the work to turn the pile of bricks into a home – this is the *efficient* cause. Then even the dodgiest cowboy builder needs a blueprint to work from: the blueprint is the *formal* cause, the plan behind the work. Finally the house must have a purpose: purpose is the *final* cause. Aristotle's point was that you need all four of these causes if the end product is to exist. Nobody would build a house if it had no purpose; without a plan the work could not be organised; and the building could not be constructed if there were no materials to build it from.

Aristotle's ideas struck a deep chord with medieval western culture. St Thomas Aquinas had realised that in Aristotle's logic there was support for a proof of the existence of God: there can be no final cause without a mind to frame the purpose. However, one of the patrons of the philosophy of modern science, Francis Bacon, who died in the seventeenth century, argued strongly that philosophy and theology should be kept separate, and that we should concentrate our studies on local problems and the interconnections between material and efficient causes. Final-cause analysis was just a distraction: 'Inquiry into final causes is sterile, and, like a virgin consecrated to God, produces nothing.'

Today we follow Aristotle's analysis if we are thinking about objects constructed *by humans* for definite purposes. All our artefacts – from hand-axes to cave paintings to inter-continental ballistic missiles – clearly have a purpose of some kind, a final cause. But modern scientists don't ascribe a conscious design to inanimate objects. We say that the planets' orbits of the sun are caused by gravity, but we don't believe those orbits are *for* anything.

In Hutton's time, though, as Maclaurin had taught him, the notions of design and final cause were far from defunct. Newton himself had been a believer in design arguments. Even a century after Hutton, Oxford geologist William Buckland would claim that coal had been put there by God for our purposes: '[the coal seams] were, ages ago, disposed in a manner so admirably adapted to the benefit of the human race.'

Harvey's work had shown that design arguments could actually lead to significant discoveries. His study of valves in the body's veins had 'invited [him] to imagine, that so Provident a cause as Nature had not so placed so many valves without Design' – the 'Design' being to circulate the blood through the systems of arteries and heart. Using such principles Harvey was able to predict the existence of capillaries, fine tubes, to close the circulatory loops; the capillaries had later been duly discovered, just as predicted. Harvey admitted, 'The authority of Aristotle has always had such weight with me that I never think of differing from him considerably.'

Against this background Hutton began his thesis, not with observations about blood, but by setting out, Maclaurin-like, Deistic attitudes about the universe. God does not intervene randomly in the affairs of the world, he argued. Nature is everywhere governed by physical, chemical and mechanical laws, and everywhere displays the wisdom and design of its Creator. Blood is a particular example of this. Blood, wrote Hutton, has been provided by God to maintain and nourish the body. But blood is used up and destroyed at the same rate, so that its supply is constant; blood orbits the body, he might have said, in as orderly a way as the planets orbit the sun (Harvey used this analogy himself). Thus nature shows evidence of rational laws and a purposeful design in the body, just as in Newton's cosmos above.

Hutton's title, meanwhile, actually harks back to another old tradition of thought, dating from the classical thinkers: that there is an analogy between the human body (the *microcosm*) and the world as a whole (the *macrocosm*).

This idea had endured throughout the Middle Ages and the Renaissance, and came into focus in the work of Leonardo da Vinci. Leonardo had developed a vision of a dynamic Earth, a living world analogous to a human body: 'We may say that the Earth has a spirit of growth, and that its flesh is the soil; its bones are the successive strata of the rocks which form the mountains; its cartilage is the tufa stone; its blood the veins of its waters.' It was a beautiful vision, and it shaped Leonardo's art as well as his science. In the *Mona Lisa* the flows of the model's hair and drapery emphasise the harmony between the processes of her body and the vivification of the watery, changing Earth.

In Hutton's own time, the unity of a living creation tied together by a 'Great Chain of Being', a hierarchy that spanned from atoms to God, was a common notion. Perhaps Hutton had read Pope, who wrote in 1733: 'Vast chain of being! which from God began, / Nature aethereal, human, angel, man, / Beast, bird, fish, insect, what no eye can see, / No glass can reach; from Infinite to thee, / From thee to nothing.'

Hutton's medical arguments were a remarkable foreshadowing of his later visions of a geocosm: of our Earth as a unified living world, designed by God to fulfil His divine purpose. In 1795 he would write, 'All the surface of this Earth is formed according to a regular system of heights and hollows, hills and valleys, rivulets and rivers, and these rivers return the waters of the atmosphere into the general mass, in like manner as the blood, returning to the heart, is conducted in the veins.' And he would come to believe that just as our bodies are capable of recovery from illness, so the Earth is capable of self-renewal: 'We are thus led to see a circulation in the matter of this globe, as a system of beautiful economy in the works of nature. This Earth, like the body of an animal, is wasted at the same time that it is repaired.'

All this in a title; Hutton did not explore these ideas in detail in his text. Perhaps, given the religious orthodoxy of the time, he did not dare say more.

Hutton was awarded his medical degree in 1749. By now his restless mind was already wandering away from medicine, and even from chemistry, his first love. For his travels had brought a new realm of mystery to his attention: the mystery of the rocks.

After his introduction to geology in the Clerks' mines, Hutton probably attended Professor François Rouelle's lectures on mineralogy in Paris. Rouelle pioneered ideas, concerning the order in which rocks had been laid down, that would later become highly significant. It is certainly possible that Hutton noted some peculiarities in the rocks of the countrysides he visited – so different from the ancient, tortured landscapes of Scotland. Forty years later, lecturing to a learned audience, he would describe the landscape of France and the Low Countries, and urged his listeners 'to examine the chalk-countries of France and England, in which the flint is found variously formed ... More particularly, I would recommend an examination of the insulated masses of stone, found in the sand-hills by the city of Brussels; a stone which is formed by the injection of flint among sand.'

Flint nodules in sandstone: it sounds an innocuous enough observation. But the trouble was, as Hutton surely realised by now from Professor Rouelle's lectures, that the primitive geological theories of the time had absolutely no explanation for how those nodules had got there.

6

'Upon this chaos rode the distressed ark'

We live on a restless Earth. Our continents are rafts floating over a hot, turbulent mantle; over geological time they drift, collide, merge and shatter.

Some two hundred and fifty million years ago, the continents began to converge into a single land mass: Pangaea, a union of all Earth's landscapes. In Scotland, compressed at the heart of a supercontinent, there was volcanism: the great cores on which Edinburgh is built are a relic of this era. Pangaea's tremendous geological unity was ephemeral, soon shattered by great oceanic rifts. The opening up of a divide between Greenland and Scotland would sunder Europe from North America for good – or at least until the time of the next supercontinental congress in the distant future. It is strange to think that the border between England and Scotland, so long the scene of bloody battles between the nations, is a genuine physical boundary – the place two continents collided – and that the Great Glen is actually a geological fault whose extension can be traced across the Atlantic to Greenland and Newfoundland.

Later, as dinosaurs hunted, the oceans rose. On the floors of the shallow seas which covered much of Britain, great layers of chalk built up. The final touches to Scotland's geological formation were made over the last two million years, as ice sheets, kilometres thick, crushed and gouged the hardest bedrock. Britain was left with one of the most varied geological landscapes in the world, from the ancient, contorted basement rocks of Scotland's far north to the soft hills of southern England – chalk hills, untouched by the glaciers, just sixty-five million years old. It is a remarkable geological story, whose outcome is a landscape of beauty and drama – and it is a story told in the rocks that remain, distorted, eroded, uplifted and shattered.

Hutton's interest in geology grew. Playfair said he 'became very fond of studying the surface of the Earth, and was looking with anxious curiosity into every pit, or ditch, or bed of a river that fell in his way'. Perhaps he glimpsed something of the titanic story of the past in the rocks he studied, and he must have longed to understand more; but he found himself illiterate in a great library of what he would call 'God's books' – and there was nobody in the world who could have taught him to read.

In his geological curiosity, though, he was in distinguished company.

Once, when Leonardo was working on his great equestrian statue to Francesco Forza in Milan, some peasants brought him a large sack of fossilised shells and corals. This wasn't so remarkable – except that the shells had been found high in the Apennine mountains. Hutton himself would later observe 'sea shells in the travelled soil a considerable height above the level of the sea' in a colliery at Kinneil, which he visited in 1765 to view James Watt's first steam engine: 'There is a bed of oyster shells some feet deep appearing in the side of the bank, about twenty or thirty feet above the level of the sea, which corresponds with old sea banks ...'

Sea creatures on mountain tops?

If the study of the Bible for clues about Earth's origins had a long tradition, a fascination with the rocks that formed Earth's surface was older still. The Mediterranean is one of the world's most active seismic and volcanic regions, and many Greek and Roman thinkers of the classical age had been intrigued by volcanoes, earthquakes and the nature of rocks. Strabo, alive at the time of Christ's birth, wondered if volcanoes were like natural safety valves, a release for the Earth's trapped vapours.

In western Europe after the Renaissance, there was a growing interest in Earth and its riches. The demand for minerals was increasing, not least because the replacement of medieval feudalism by capitalism created a need for various precious metals for coinage. This led to new and more pragmatic geological works based on what was actually found in the ground. For example, Georgius Agricola's *De Re Metallica*, published in Basle in 1556, gave a fine discussion of the occurrence of mineral veins in the field, and tips on practical geology.

From the beginning, though, fossils like those of Leonardo and Hutton had posed special puzzles.

In the fifth century BC, Herodotus had noticed marine shells far inland in Egypt, and had speculated that they had been left there by the retreating waves of some earlier sea. Earlier still, Pythagoras had wondered if fossils observed high in the Greek mountains could be evidence of the elevation of an old sea bed. In Leonardo's time, fossils were called 'figured stones'. In France, Italy, Germany and America, people gaped at huge bones and tremendous teeth thought to have come from the drowned corpses of the 'giants' mentioned in the Bible. (Many of these relics eventually turned out to be from the vanished fauna of the Ice Age: mammoths and mastodons, cave bears and woolly rhinos.)

But for the scholars, the most baffling questions about fossils were much more subtle. These awkward articles looked like nothing so much as copies, in stone, of the relics of living things: bones, and the teeth of sharks, and the shells of sea creatures. What *were* they – and how had they got inside solid rock?

The similarity of the fossils to living counterparts, so obvious to modern eyes, wasn't seen as a clinching proof of their organic origin. God was thought to have imbued His creation with similarities on different scales to display the harmony of His thoughts. Thus the existence of seven 'planets' (the sun, Moon and the five planets known to the ancients) was in accordance with the seven notes of a musical scale. Perhaps God had simply granted rocks with the ability to form objects exactly like animal parts. Why not? Everything was part of God's narrative, to be written as He wished.

But with arguments like that, you could explain away literally anything – in the nineteenth century Charles Lyell would observe in such theories 'a desire manifested to cut, rather than patiently untie, the Gordian knot' – and anyhow such justifications weren't much practical use in figuring out where the best mineral veins were. From these and other dissatisfactions, a new way of thinking about the Earth began to emerge.

Nicolaus Steno had come to reside in Tuscany. A Catholic convert, in 1677 he would be appointed the Titular Bishop of Titiopolis (now part of modern Turkey). The see of a titular bishop covered areas *in partibus infidelium* – in the hands of unbelievers, so not available for actual occupancy – but Steno would earn his living by ministering to those scattered Catholic communities that clung on in post-Reformation Germany, Norway and Denmark.

Like Hutton, Steno had a medical background. He was an anatomist, and his ordination actually led to him giving up his science. His contribution to geology – monumental in retrospect – came almost as an afterthought to his secular career, just before his energies were elevated to higher things.

In 1666, Steno was sent the head of a great shark, caught off the coast near the town of Livorno. Steno was already aware of *glossipetrae* – 'tongue stones', fossils shaped oddly like sharks' teeth that could be collected by the barrel-load, especially in Malta. To Steno the anatomist, now able to compare the fossils with modern specimens, it was undeniable that these trinkets really did look like sharks' teeth – they were identical in form, in fact. But the old questions remained: how could sharks' teeth get *inside* the rocks?

Steno tried to take some of the arbitrariness out of geology. He followed Descartes's thinking that all phenomena must have physical causes: even fossilised sharks' teeth embedded in rocks must have a rational explanation. Steno set out rules of geological thinking. First he stated a principle of *similarity*: 'If a solid substance is in every way like another solid substance, not only as regards the conditions of its surface, but also as regards the inner arrangement of parts and particles, it will also be like it as regards the manner and place of production.' If it looks like a shark's tooth – if you take it apart and, to every level you can examine it, it still looks like a shark's tooth – then it is, or at least was, a shark's tooth.

Second, Steno set out a principle of *moulding*. If you find one object inside another, you can tell which formed first by seeing which has left impressions on the other. A fossilised shark's tooth leaves an impression on the hardened sediment that encases it, like a footprint in wet sand, so the shark's tooth must have been there first.

Steno's introduction of orderly and systematic thinking is rightly seen as a foundation stone of modern geology. The application of Steno's intellect to purely religious matters in later years has to be seen as a loss. But Steno's principles caused much sharp intaking of breath among his contemporaries. For one thing, he was limiting the actions of God, by arguing that He did not arbitrarily allow rocks to make copies of sharks' teeth.

And, Steno said, *not everything had been made all at once*. First the shark's tooth fell to the bottom of the sea, then the sediments closed over it, then they consolidated into rock ... Earth was a dynamic object; though perhaps only six thousand or so years old, Earth had *changed* in that time. Suddenly Earth had a history, and you could reconstruct that history from the rocks. For many thinkers Steno's arguments were electrifying.

We know that Hutton, his geological curiosity deepening, read some of the speculative accounts which followed Steno's work of how the Earth might have come to be – but he found that in all these works geology was still seen as 'the handmaiden of the Bible', the great book where the core truths of the universe were to be found.

The Reverend Thomas Burnet, Anglican clergyman, was a contemporary of Newton at Cambridge. One day he would become the private chaplain of King William III. He was drawn to geological mysteries after a visit to the Alps. Jagged, twisted and broken, the mountains seemed to Burnet like signs of a 'World lying in its Rubbish'. Surely God would have created a *perfect* world, not this heap of rubble. But if mountains weren't part of the original world, how had they got there?

Burnet's musing on these questions led to his publishing, between 1680 and 1690, the four books of his *Telluris Theoria Sacra* – or, to give it its full and unassuming title, *The Sacred Theory of the Earth: Containing an Account of the Original of the Earth, and of the General Changes which it has Already Undergone, or it is to Undergo, Till the Consummation of All Things*. Hutton read this avidly.

The Reverend Burnet's programme was to start with the Bible story as a given. 'In the beginning God created the heavens and the Earth. And the Earth was without form, and void; and darkness was upon the face of the deep. And the spirit of God moved upon the face of the waters. And God said, Let there be light: and there was light ...' The ancient words of the Book of Genesis surely still thrill even non-believers. For Burnet's generation, the goal of natural scientists was to seek physical causes to justify the sacred story: essentially you had to invent a physics to explain away the Bible's events, from the Creation to the present day. (In modern times, Immanuel Velikovsky has pursued a similar programme based on the texts of ancient civilisations.)

But in fulfilling this programme Burnet was determined to cling to Descartes's principles of the steady operations of natural forces. God made the world right the first time, and left it to run by itself. To imagine otherwise is to belittle Him: 'We think him a better artist that makes a clock that strikes regularly at every hour from the springs and wheel which he puts in the work, than he that hath so made his clock that he must put his finger to it every hour to make it strike.' Burnet's task was to reconstruct the orderly operation of God's great terrestrial clock.

The centrepiece of Burnet's tale was Noah's Flood. As the most spectacular geological event since the Creation itself, perhaps the Flood could be invoked as the agent that had killed off all the creatures whose relics were now being found in the rocks. And perhaps, further, the rocks that contained those relics had themselves been precipitated out of a universal Flood-created sea.

Burnet tried to figure out where enough water to cover the mountains could have come from. There wasn't enough water in the oceans – even forty days and nights of rain would add only a metaphorical drop – and so he concluded that there must have been more oceans lying deep underground, a worldwide layer of water under Earth's original crust. This gave him the clue that allowed him to construct his cosmology.

Burnet's primordial universe was a jumble of particles from which Earth coalesced in a smooth series of concentric layers, sorted by density. The body of the planet contained a vast layer of water. Earth's original crust, lying over the water, was perfect, its surface featureless and smooth: 'no Rocks nor Mountains, no hollow Caves, nor gaping Channels, but even and uniform all over'. Rivers flowed from the poles to the tropics, where they dissipated. Earth's axis had no tilt, so there were no seasons. Eden, placed at a mid-latitude, enjoyed perpetual spring, and everybody lived to nine hundred years or more.

The Flood occurred when this original crust cracked open. As the Bible tells us, the 'fountains of the great deep' erupted. The unfortunate planet tipped over (perhaps pushed by angels, as in Milton's *Paradise Lost*), and the endless spring was lost. In these unpleasant conditions lifespans shrivelled to a mere three score years and ten. The Earth's current surface is nothing but a ruin of what went before the Flood. The ocean basins are gaping holes, the mountains upturned fragments of the old Edenic crust. Since then geological processes have only served further to erode this global wreck. Only six thousand years old, Earth has already fallen into old age.

But these unhappy times won't last forever. In future ages a new deluge will come – but this time a deluge of fire, sparked by volcanic torches and fed by huge underground reservoirs of air. (Britain's coal reserves will help it burn particularly brightly, Burnet adds on a cheery patriotic note.) This conflagration will consume the Earth, mobilising its particles into a new period of chaos. But these particles will settle out once more into a new perfection, with concentric layers sorted by density. On an Earth made perfect again, Christ will reign for a thousand years. Finally, after a final battle against the forces of evil, the saints will ascend. The Earth, abandoned, will become a star.

In the telling, Burnet showed a mastery of divine special effects: 'Upon this chaos rode the distressed ark, that bore the small remains of mankind. No sea was ever so tumultuous as this ... The ark was really carried to the tops of the highest mountains, and into the places of the clouds, and thrown down again into the deepest gulfs.' Burnet's account was a popular hit: he was the Cecil B. de Mille of Mosaic geology.

We have to embrace the progress of understanding, but I think we are allowed to mourn the loss of such magnificent stories. That's not to say, however, that the story of the Earth as we know it today doesn't contain its own drama. Indeed, it even exhibits strange echoes of Burnet's busy cataclysms.

Now we believe that the Earth was indeed formed from a chaotic cloud, a lenticular swarm of dust and gas that orbited the young sun. And we do believe that the world's surface will be destroyed by fire – in the far future of a billion years or more, when the ageing sun will swell into a 'red giant', its surface creeping like a crimson tide past the inner planets, at last breaking upon the orbit of the Earth. Today we no longer imagine that Earth will become a star, but it will be destroyed by one.

Burnet's work left Hutton unsatisfied, however, for all its drama. Without a solid logical foundation – without some way for Burnet to show *how* he had constructed his vision of the past, which after all we cannot observe directly – it remained just a story, if a marvellous one. Hutton would write that 'it surely cannot be considered in any other light than as a dream, formed upon a poetic fiction of a golden age.'

More theological system-builders followed Burnet. But baffling problems remained.

The notion that the rocks had been deposited out of a universal ocean made a certain sense. It could even explain the strata to be observed in many of the rocks, if the particles in the ocean were deposited out in an orderly way.

But the strata were not always found stacked in neat layers. Sometimes – in Scotland, usually, in fact – they were uplifted, broken and folded. They can even be found folded back on themselves altogether. And then there were the peculiarities to be found within the strata, like Hutton's flint nodules, and veins of minerals and basalt that cut *across* the strata of other rocks. It was hard to imagine any sequence of events in a drying ocean that could have caused such anomalies.

And then there was time.

This troubled Burnet himself. All the marvellous events of the Creation seemed to Burnet to require rather more than the standard six days of the Genesis story to achieve. Perhaps, he mused, the six days were allegorical. He conducted a correspondence with Newton over this, but Newton preferred the idea that there really had been six days, but that the early 'days' had been of indeterminate length.

If the age of the Earth was in question, however, perhaps it would be possible, not just to deduce it from the Bible – but to *measure* it.

Georges-Louis Leclerc de Buffon had a colourful life. A Frenchman, he was born into riches, and never lost the taste for the good life. Buffon was, however, a dutiful scholar, and the works of Newton made a profound impression on him, as on so many others. In 1734, aged twenty-six, he was elected to Paris's Academy of Sciences, and five years later became director of the *Jardin du Roi* – the Garden of the King, a botanical garden that became one of France's leading scientific institutions.

In a quest for scientific glory, Buffon turned his formidable energies on the question of Earth's origins. Buffon was different from many contemporary thinkers in that he rejected the Bible story. Newton's laws had been sufficient to explain the motion of the planets and the oscillation of the tides; perhaps they could be applied to the history of the world itself. Buffon formulated a mighty new theory. He proposed that an immense comet had splashed into the sun: the impact had thrown off a great gout of molten material that had coalesced into the planets.

It was just another 'poetic fiction' – but Buffon used these ideas to produce an estimate of the age of the Earth. A globe of iron an inch in diameter would cool from red-hot to room temperature in an hour, larger globes longer. So how long would it take to cool a globe of iron the size of the Earth? Buffon's result, worked out from experimentation and calculation, was startling: some 74,000 years – much longer than the Ussher chronology.

Buffon's work created a great furore, of course. The theologians at the Sorbonne condemned him, and Buffon dutifully retracted. But he wasn't sincere: 'It is better to be humble than hanged.'

In any case, privately Buffon had begun to believe that even 74,000 years was *still* too short. He continued with private experiments, and came up with new estimates of Earth's age: a million years, three million, ten million. He didn't publish such numbers, not so much from fear of the religious authorities – by now he was too old to care – but because he thought the public wasn't ready for them: 'Why does the human mind seem to lose itself in the length of time?'

Six thousand years, seventy-four thousand, a million years,

ten million ... The predictions of Newtonian natural law and the evidence of the rocks – including such awkward anomalies as Leonardo's fossils and Hutton's flints in sandstone – were becoming increasingly difficult to marry to the Ussher chronology. But there was no alternative theoretical framework to deal with these issues, no other way to think. If the world hadn't been created as the Bible set out, nobody had much of an idea how it actually *had* been.

Hutton, distracted by the puzzles of the rocks, was surely intrigued by this slowly gathering debate – but he had no time to take part in it, for once again his young life was to be plunged into turmoil.

7

'The wandering infidelities
of the heart'

In 1749, having qualified as a doctor at the age of twenty-three,
– Hutton returned to Britain. He lingered in London through
the winter, and then travelled back to Scotland. He would
spend two years in Edinburgh, and the fallout from this time
would affect the rest of his life. For they were years in which
he wrestled with indecision over his future – and in which he
had a disastrous love affair.

Hutton certainly didn't want to be a doctor. He was still
devoted to chemistry, the subject that had drawn him into
medicine in the first place, but he had become cynical about
contemporary medical science – and he didn't have a
physician's nurturing instincts. Much later, his friend Adam
Ferguson would say that 'an attempt to consult him or see him
[over a medical matter] would have been met with a laugh, or
some ludicrous fancy, to turn off the subject.' Besides, the
business of medicine in Edinburgh had been sewn up by a few
long-established practitioners: there was no opening for a
young man without connections, recommendations or track
record – or, as Playfair put it, any of 'that patient and
circumspect activity by which a man pushes himself forward
in the world'. And work as a physician would have made
relentless demands, leaving him little time to pursue
chemistry, geology and other interests.

So Hutton started to explore an alternate path. One of his
old friends, of about his age, was James Davie. Their relationship was based on a common interest in chemistry, and before
Hutton had left Edinburgh for his foreign medical training
they had dabbled with experiments in the production of sal
ammoniac – that is, the salt of ammonia and hydrogen
chloride. In natural form this industrial chemical is a white
crystalline salt. It is used today as an electrolyte (a conducting
fluid) in small electrical cells of the type used in torches and

radios, and in many cough medicines and cold remedies. In Hutton's day it was important for dyeing, and for working with brass and tin. Davie had continued the experiments in Hutton's absence, and had come up with a way of manufacturing the salt from common fireplace soot, a resource readily available from the hearths of Edinburgh. Since the salt had previously been available only as an import from Egypt, this was an obviously profitable prospect. Hutton and his partner began to develop a works to manufacture the salt. Davie took the lead at first, and the factory was established some time during the 1750s.

It was an unglamorous business. Davie signed a deal to take all the soot collected by the 'tronmen', the Edinburgh chimney sweeps. To process the soot, glass spheres about thirty centimetres across were filled almost to the brim with the stuff. Davie obtained the spheres from a glassworks in Prestonpans run by John Roebuck, who would later partner James Watt. The spheres were put into a furnace and heated up, with their mouths outside in the cooler air. The heat drove the sal ammoniac out of the soot, and the salt would be found to have collected on the spheres' cooler surfaces around the mouths, in thick lumps about five centimetres deep. In this way you could reduce ten kilograms of soot to yield three of sal ammoniac.

It must have given Hutton great satisfaction that his chemical tinkering, which had deflected his university career and caused him to lose his job as a legal apprentice, should now turn up trumps; his restless mind and mischievous nature were at last paying dividends. But it would take a while for the sal ammoniac business to turn a reasonable profit, and Hutton still had to work for a living. Again he cast around for a new avenue.

But now his life was devastated by scandal.

Though Hutton struck Playfair as an open character, he was actually very reticent about his personal life. He never told the full story of these years, even to his closest friends. But there are hints of what happened from autobiographical fragments in his later works, and from his few surviving letters.

The world knew him as a bachelor, but in his letters Hutton hinted he was trapped in a relationship – and perhaps he was even married: 'Now I am e'en wedded, and so must endeavour to restrain the wandering infidelities of the heart.' Hutton seems to have thought the affair had ruined his life, and became bitter towards women in general, though he himself seems not to have been without fault: 'I don't let any of the fair kind of creatures know of my distress it would kittle the malicious corner of their hearts to hear the afflictions of a hardened wretch whom they could never make to groan.'

Surely it wasn't Hutton's first love affair. But this one went sour. What we do know is that it produced a son, born around 1749. Hutton seems to have shut the child out of his life; he never told his friends of the boy's existence. But the birth of the child caused Hutton great distress. He appears to have been forced out of Edinburgh under duress, presumably because of the scandal.

He still had to make a living, though. Exiled from the city, he had just one fall-back position: the two farms bequeathed him by his father, at Slighhouses and Nether Monynut in Berwickshire.

James Hutton – a *farmer*?

Playfair, knowing nothing of the scandal, would try to rationalise Hutton's choice: 'We ought ... I think, to look for the motives that influenced him, in the simplicity of his character, and the moderation of his views, [rather] than in external circumstances. To one who, in the maturity of understanding, has leisure to look around on the various employments which exercise the skill and industry of man, if his mind is independent and unambitious, and if he has no sacrifice to make to vanity or avarice, the profession of a farmer may seem fairly entitled to a preference above all others ...'

But this bucolic diagnosis is mistaken. Hutton must have viewed the prospect with horror. Slighhouses was not exactly a comfortable estate. It was a small, remote farm that would be difficult to work, and where a city boy would be isolated, and starved of entertainment and intellectual companionship. However, by 1752, three years after completing his medical degree, Hutton decided at last that he had no choice.

It wasn't necessarily a bad time to become a farmer: Scottish agriculture was about to undergo a revolution.

In the harsh climate of the early years of the Union, many
Scottish gentry had become embarrassed by the comparison
between their poor farms and their English counterparts. So in
1723, three hundred patriotic types formed the Honourable
Society of Improvers in the Knowledge of Agriculture. The
Improvers experimented with foreign techniques of crop
rotation and enclosures, new tools, ploughs and milling
machinery, and they reaped their rewards in increased yields
and profits. Cattle began to be moved on drove-roads that
spanned the Highlands, and the cultivation of the potato in the
Lowlands would provide a reliable diet for the ordinary folk for
the next hundred years.

Eventually Scotland's farms would become models of
agricultural science, admired and copied worldwide; the
Improvers' movement was one of the first flowerings of the
great Scottish Enlightenment that would characterise the rest
of the eighteenth century. Of course there were human costs:
common pastures were fenced off, small farms combined into
larger holdings, and unwanted tenants evicted. However,
resistance was dispirited and disorganised, and ended with the
transportation of some of its leaders.

Hutton was aware of the new techniques. 'It was, I believe,'
he would write, 'by reading the ingenious Mr [Jethro] Tull at an
early period of my life that I acquired a taste for agriculture.'
Besides, he was determined to put off the dread prospect of
actually breaking the soil as long as possible. Having been
thrown out of Edinburgh, Hutton didn't make immediately for
his farms, but travelled south, on the not unreasonable pretext
of studying a little agricultural science first.

In Edinburgh he had become acquainted with a Norfolk
gentleman called John Manning. At the time, Norfolk led the
country in improved methods of agriculture: the land there
had been enclosed, and a fourfold rotation system, of roots,
barley, clover and wheat, had led to vastly better yields of both
animals and crops. So for a time Hutton stayed with Manning's
father in Yarmouth. Later he stayed with another gentleman
farmer, John Dybold of Belton, who served as 'both his
landlord and instructor'. Hutton lived in the farmer's house,
and was put to work, enjoying many 'practical lessons in
husbandry'.

Hutton seems to have enjoyed Norfolk. Playfair wrote, 'The simple and plain character of the society with which [Hutton] mingled, suited well with his own, and the peasants of Norfolk would find nothing in the stranger to set them at a distance from him ... There was accordingly no period in his life to which he more frequently alluded, in conversation with his friends; often describing, with singular vivacity, the rural sports and little adventures, which, in the intervals of labour, formed the amusement of their society.'

After a year with Dybold, Hutton moved to central Suffolk, where he compared farming on heavy land with techniques suitable for the light Norfolk soils. He made studies of ploughing, and of dairying and butter manufacture.

And during these two East Anglian years, Hutton explored.

He made many journeys – mostly on foot – to different parts of England. He visited Northumberland, Yorkshire, Derbyshire, Cambridgeshire, Oxfordshire and the Isle of Wight. These trips were made primarily to study agricultural techniques, but by this time Hutton had also begun to study geology and mineralogy in a more serious way.

Playfair imagined that he took up the subject more deeply in order to 'amuse himself on the road'. If you were a geologist, there was, after all, always something to see, and on horseback or on foot Hutton was immersed in the countryside. Perhaps, too, as with chemistry, he was attracted by geology's tactile immediacy: you could touch the rocks, turn them over, explore them with your hands as well as your mind.

Soon he started to direct his journeys to interesting geological features, as well as to agricultural objectives. Hutton would later boast that he could tell you where a piece of gravel had come from anywhere on the eastern side of Britain.

He followed, as he would write in 1770, the 'ridge of indurated chalk that runs east and west thro' the Isle of Wight, the Needles are portions of this remaining still undemolished by the waves, it seems to be continued, under the sea, west into the opposite coast where needles are also formed of it, and I travelled upon the top of this ridge from Corfe Castle to Weymouth.' And as Hutton explored these soft young rocks of southern England – so different from the ancient and heavily glaciated rocks of Scotland – what he saw struck him greatly.

Playfair, later desiring to 'trace the progress of an author's mind in the formation of a system where so many new and enlarged views of nature occur', gives us some hints of Hutton's thinking at this point in his life. 'It appears that ... [the proposition that] a vast proportion of the present rocks is composed of materials afforded by the destruction of bodies animal, vegetable and mineral, of more ancient formation, is the first conclusion that he drew from his observations. The second seems to have been that all the present rocks are without exception going to decay, and their materials descending into the ocean. These two propositions, which are the extreme points, as it were, of his system, appear, as to the order in which they became known, to precede all the rest.'

At Yarmouth, Hutton saw a spectacular example of Earth's destructive forces at work. Where the River Yare meets the North Sea, Hutton watched a flood carry away part of the land, and storms batter the coast. Hutton, intrigued, sought out more examples of erosion, like the Needles of the Isle of Wight. Erosion was everywhere shaping the world, and it was relentless.

Hutton knew about the practical implications of erosion, of course. The Norfolk farmers stressed the dangers of soil erosion and depletion, and how to counter it with crop rotation, good fertiliser and the right kind of drainage ditches. But as a trainee farmer he must also have observed that erosion isn't always a bad thing. All rocks are continually eroding away, as wind and water, heat and frost work at them. But then nothing can live on bare rock. Erosion might be destructive – but it is actually *necessary*, to make the soil that is the substrate of all life. It was a paradox.

And if the land was being destroyed, Hutton also noticed how much of the rock exposed along the Yarmouth coast showed evidence of recent *creation*. Some sandstones looked just like tightly packed sand – it was just that the grains had been somehow cemented together – and other rocks were full of fossil shells that had obviously once, and comparatively recently, been under the sea. 'Dr Hutton,' Playfair would write, '... insists much on the perfect agreement of the structure of the beds of grit and sandstone, with that of the banks of unconsolidated sand now formed on our shores, and shews that these bodies differ from one another in nothing but their compactness and induration.' In a way this was an application of Steno's principle of similarity.

If river and rain and ocean could destroy rocks, there were clearly forces working in the world that could *make* them again, and even lift them out of the oceans where they were formed. But though the young science of geology might be the 'handmaiden of the Bible', this creativity had nothing to do with scripture. The Bible does describe erosion: Isaiah tells how 'every valley shall be exalted, and every mountain and hill shall be made low; and the crooked shall be made straight, and the rough places plain.' But scripture contains no hint of how the land might be *restored*. So erosion is a one-way street towards destruction and uniformity. Whatever world-building system you adopted, if you accepted the Bible story the world *must* be young, because otherwise erosion would already have destroyed all its topography. In studying sandstone, however, Hutton had clearly perceived that creative forces did exist – even if the Bible didn't say so.

In the spring of 1754, Hutton decided to visit the Low Countries once more, as that was where most of the recent English innovations in agriculture had sprung from. So he set out from Rotterdam and travelled through Holland, Brabant, Flanders and Picardy. He was impressed with what he saw, but made a comparison flattering to England: 'Had I doubted of it before I set out,' he wrote, 'I should have returned fully convinced that they are good husbandmen in Norfolk.'

He took the chance to study again the chalk and sandstone formations he had first noticed during his days as a medical student. Chalk was relevant to his farming studies, because it was used as fertiliser throughout southern England. But here was another puzzle. Chalk was a classic sedimentary rock, clearly created from the detritus of earlier erosion, and from the bodies of sea animals: 'That which renders the original of our land clear and evident,' he would write in 1785, 'is the immense quantities of calcareous bodies which had belonged to animals, and the intimate connection of these masses of animal production with the other strata of the land.' Yet this new rock, so obviously created under the sea, was essential for life.

So erosion destroyed rock, unknown forces created it, and in the midst of all this the soil necessary to support life was created. It was all fascinating – but not explained by any geological theory then extant – and certainly, to Hutton's mind, none of it fitted the Bible's traditional picture of a world in decay. All these observations swirled in his mind, pieces of a puzzle that lacked any obvious key.

This interval in England and Holland was a relatively happy time for Hutton, full of studying, socialising, travelling and geologising. Perhaps it was a welcome break from the hothouse crisis he had left behind in Edinburgh. But by the summer of 1754, after two years away from home, he couldn't put off the black day any longer. It was time to get his hands dirty.

8

'A cursed country where one has to shape everything out of a block'

The land was wild and uncultivated, just open fields that backed on to sheep country. Stones had to be split and hauled away before Hutton could work the soil at all. Much of the farm was sculpted into long ridges intended to help drainage, but whose principal effect was to wash away the topsoil. Hutton had to flatten the old ridges, and dig new drainage ditches.

It was slow, back-breaking, exhausting work. Hutton found it hard even to find tools and skilled tradesmen. 'Talk of moving mountains, I can tell you, I'm obliged to use a great clumsy wooden slipe & drag the whinstone through the rough land (sore against their will) ... my houses are not a foot more forward than the day I came to this country.' He was stranded in 'a cursed country where one has to shape everything out of a block & to block everything out of a rock ... I find myself already more than half transformed into a brute.'

But he persevered.

Hutton's farms are six kilometres north of Duns in Berwickshire, about sixty kilometres south-east of Edinburgh. Slighhouses had been bought by Hutton's great-uncle John in 1713. John sold it to three of his nephews, including Hutton's father William; one of the brothers died, and William bought out the third. Meanwhile William had bought Nether Monynut outright in 1710. ('Nether' means 'lower'. There is also a Middle Monynut and an Upper Monynut, making a string of holdings along the Monynut Waters which tumble down from the Lammermuirs.) In 1760, Hutton would have to go through a legal procedure to re-establish his ownership of the farms.

To get to Slighhouses you travel south from Dunbar through the arable country of the Lammermuir Hills. Today this is a gently rolling landscape of dry stone walls, sheep and cattle.

Slighhouses itself is some hundred metres above sea level. To the east the land falls away, affording a long and pleasing view towards the Tweed valley, while to the west the land rises towards the Lammermuirs. Slighhouses is unremarkable and mostly modern, though its buildings have eighteenth-century rooms built onto an older core – perhaps improvements made by Hutton. Meanwhile Slighhouses' sister farm, Nether Monynut, is two hundred metres higher, and the soil is thick and stony. It isn't clear what use Hutton made of the hill farm, but perhaps his father had used it as a source of stock for his arable land.

It is difficult to imagine how it must have been when Hutton first arrived here, and how he must have felt about his sour personal circumstances. But despite the unpromising location, Hutton was determined to apply the modern methods he had learned in the south, and make a success of his farming. He probably applied a crop-rotation system, growing wheat in the first year of the cycle, then turnips, and finally barley, undersown with grasses and clover to provide pasture for the following year. He kept cattle and probably sheep.

His most notorious innovation was to import modern ploughs: light and small, and drawn by only two horses. Adam Ferguson would write that in Norfolk Hutton 'purchased a plough, hired a ploughman, and brought both on the post-chaise with him to Berwickshire. The neighbours were diverted with this assortment of company and baggage, and no less with the attempt which followed, to plough with a pair of horses without a driver.' This vignette is probably not completely accurate; Hutton brought several ploughs with him from East Anglia, and when he was unable to find a local willing to learn how to use them, he was forced to go back to Norfolk to persuade an experienced ploughman to come to Scotland.

Starved of companionship, Hutton took a certain sentimental pride in the cattle under his care, especially in the ancient cycle of birth and death. He wrote to his friends, 'Give me joy, this day is born into my family a male child – and the mother is in a fair way of recovery – O if the ladies were but capable of loving us men with half the affection that I have towards the cows and calfies that happen to be under my nurture and admonition, what a happy world we should have!' He learned to make butter with cream that had been scalded to get rid of the taste of turnip, and he became something of an expert on cheeses.

Nonetheless, living alone, Hutton was soon lonely, bored and depressed: 'This squeamish homebred stomach of mine isn't truly reconciled to the bitter pill of disappointment.' Despite his social success in East Anglia, the life of the Berwickshire rustics was too uncouth for his tastes: 'They had me at a feast of Baal where was a sow an honest sow roasted i' the guts so we had a dish of surprised pig and I did eat thereof, they led me up into the dance, but I will enter no more into their high places.' He missed city life, and tried to persuade his friends to visit him. Later, he would write of his loneliness during a solo geological tour: he would 'walk about & enquire to prevent hanging himself through the day and then at night he writes a bit and drinks a sup of hot toddy.'

His neighbours were a comfort, however. During his time in Edinburgh Hutton had met Sir John Hall of Dunglass. Hall, fifteen years older than Hutton, was a man 'of ingenuity and taste for science, and also much conversant with the management of country affairs'. Dunglass is only some thirteen kilometres from Slighhouses, and Hall and Hutton would remain close friends. The friendship would continue to the next generation: Hall's son James would accompany Hutton on that fateful trip to Siccar Point three decades later.

David Hume's older brother James was another of the agricultural Improvers. The Hume family home was at Ninewells, no more than five kilometres from Slighhouses, and it is possible that Hutton made David Hume's acquaintance during this period.

The Clerks remained companions too. As the seventh of his father's sons, Hutton's old friend from medical school, John Clerk of Eldin, was never wealthy. In 1762, he purchased a small coalfield from the Marquis of Lothian. To raise the money he took out bank loans, and received help from 'our most benevolent and worthy friend Doctor Hutton'. Playfair described Hutton as always 'humane and charitable' to his friends: 'He set no great value on money, or, perhaps, to speak properly, he set on it no more than its true value.' Unable to afford a manager, Clerk had to direct the mining operations himself – and the extensive knowledge of sedimentary rocks and their structure he acquired must have been of great interest to Hutton.

The geology, indeed, was always a consolation.

In 1764, after ten years at Slighhouses, Hutton took a break from his farming. He had remained a bachelor, and probably left his farm in the hands of his Norfolk ploughman while he set off with George Clerk-Maxwell to tour the Highlands.

George – later Sir George, fourth Baronet of Penicuik – had added 'Maxwell' to his name after marrying his cousin Dorothea Clerk-Maxwell. He had developed interests in mining, manufacturing and agriculture, and wrote learned pamphlets on such subjects as fisheries and ploughing. Hutton had become great friends with Clerk-Maxwell, of whom Playfair said, 'a gentleman distinguished by his abilities and worth, with whom Dr Hutton had the happiness to live in habits of the most intimate friendship'. Clerk-Maxwell was the first, but not the last, of the Clerk brothers to accompany Hutton on significant excursions into the field.

Hutton and Clerk-Maxwell travelled from Crieff, near Perth, over the Grampians to Dalwhinnie. They made their way to the Great Glen, travelling from Fort Augustus along the shore of Loch Ness to Inverness. Then they crossed to Aberdeen and travelled down Scotland's east coast, eventually heading back towards Edinburgh.

Hutton was by now in his late thirties, and this tour was gruelling, if rewarding. But he could not have ignored the fact that he was travelling through a land effectively under military occupation.

After the immediate terror that had accompanied the putting-down of the 'Forty-Five, the clan system was attacked with the law. The powers of the chiefs were removed, and prohibitions were imposed against the carrying of weapons of war. Even Highland dress and the Gaelic language were banned. Meanwhile a rigorous martial control was enforced throughout the Highlands. Great military highways had been driven across the central massif, incidentally providing comfortable routes for Hutton and Clerk-Maxwell. A formidable artillery fortress was built at Fort George, east of Inverness. This huge fort is still in use as a military base today; it is a monument to the Hanoverian government's determination that the Highlands would never rise again.

All this was the beginning of the process that would lead to the Clearances, a devastating transformation that would take a century to complete, with the people of the glens being replaced by vast herds of sheep – often owned by the old clan chiefs, in a final betrayal of their 'families'. The Clearances were so thorough that during the Russian war of the 1850s – more than a century after the departure of the Bonnie Prince – when the call came to the Highlands to raise regiments to replace those destroyed at Balaclava and elsewhere, barely a platoon could be raised: 'Since you have preferred sheep to men,' growled a local, 'let sheep defend you.'

Hutton was the son of a merchant, and now a gentleman-farmer. Most Lowlanders (with honourable exceptions, such as David Hume) had simply been relieved that the disaster of the 'Forty-Five had been averted, and the clans controlled. Indeed George Clerk-Maxwell was Commissioner for the Forfeited and Annexed Estates: Clerk-Maxwell's own purpose on this Highland tour was to gather information on the agricultural and mineralogical potential of lands forfeited by nobles who had supported Bonnie Prince Charlie. In common with most of his class, Hutton probably had little sympathy for the suffering of the 'wyld men' he glimpsed in the glens.

Besides, Hutton's interest was in the rocks, not the people.

He noticed Old Red Sandstone exposed at the coast at Caithness. At the Firth of Cromarty he saw 'masses ... of shells in banks many feet above the level of the present sea mark', a puzzle to match Leonardo's mountain-top oyster shells. He saw examples of rocks that would inform his later thinking, including granite on the coast near Aberdeen, and basalt, a volcanic rock that he saw at Crieff and elsewhere, apparently injected in great veins into pre-existing strata.

Whereas Hutton's early tours had been designed primarily to gather agricultural information, with geology as something of

a hobby, by now geology had replaced agriculture as his
principal interest – though he did call on a Captain Lockhart of
Balnagown, reputedly the best farmer in the country, to advise
him to use sea shells as fertiliser. Rather than trawling for data
at random, Hutton was already focusing his geologising on
gathering specific information. He had begun to compile notes
on his observations and reflections.

He had also begun a collection of geological specimens,
some of which he collected himself, but others of which he
acquired from a growing network of fellow enthusiasts. Not all
these collectors were of high quality: Hutton would write in
about 1770 of how he had commissioned 'gentlemen of my
acquaintance in different parts of the country to send me
specimens of their limestones at hazard, for they cannot
choose proper specimens; this I have done to several parts, but
hitherto either received nothing at all or nothing worth your
looking at; and the collecting them myself is a work of time
and precarious.'

He would eventually call his specimens 'God's Books' – and,
Adam Ferguson would say, he 'treated the books of men with
comparative neglect'. He did consume geological facts,
though, which he obtained by devouring books of travel and
description. In his reliance on evidence, preferably obtained at
first hand, Hutton was showing the way to the geological
methods of the future.

Hutton's geological expertise eventually became highly
developed. In letters written around 1770 to John Strange – a
friend of a friend, and a fellow student of the rocks – Hutton
gave a long and detailed description of the geology of Scotland,
as he had learned it: 'The strata in this country are much more
irregular and mixed with other masses, in which no form of
strata is to be observed, than what I have observed in England,
from whence I except Cornwall and Wales, in neither of which
I have been. We have no chalk in Scotland, so far as I can learn;
I saw a little limestone mixed in flint, that in some measure
imitated it, a little south of Dunrobin Castle.' And so on.
Hutton's descriptions are precise, substantially correct, and
clearly based on first-hand study.

It was a remarkable achievement. When he began to explore
the geology of his native land, Hutton had far less prior

knowledge of it than Neil Armstrong had of the Moon when he first stepped out of his Apollo lander. Eventually, more or less self-taught, and after years of careful observation, Hutton would build up a synthesis of the distribution and characteristics of many of Britain's principal rock formations, which in broad outline still stands up today.

Hutton was incidentally remarkably friendly and open to Strange, a man he had never even met: 'There is our Warehouse – look about you; please your own fancy; I shall hand you down any piece of stuff you desire.' To Hutton, like others of the Enlightenment generations, knowledge was something to share. As in much of Hutton's life, there is a sense of community, of mutual support. Naturalist Thomas Pennant, writing to another geologist in 1769, described Hutton himself as 'the greatest enthusiast I ever met in your way; very lively, very ingenious'.

Hutton had also developed a great enthusiasm for his native Scotland, despite the rigours of such tours. He would say to Strange, 'I will venture to say that whatever satisfaction you may have received in your examinations of England, could your health permit you to visit this Country, it would also afford further confirmation; but though it should afford you many interesting facts, I must, in justice, warn you to expect to meet with greater stumbling blocks; not only in respect to the accommodation of your person (which a philosopher in health, at least, could get over or go round) but also with regard to the penetration of your spirit; the first may deter an invalid, the second will incite a true Philosopher.'

After the tour, though, Hutton had to return to the solitude of the farm. As he would write to Clerk-Maxwell ten years later, 'If there is to be an excess I think I would choose it to lie upon the side of quietness & silence only that we shall see enough of it in the grave before the resurrection begins mayhap, so let us speak a little before we lie down.'

Hutton persevered with his farming, and slowly his situation improved. His Norfolk plough transformed the land. As Adam Ferguson noted, 'The joke [of his Norfolk plough] has become serious, and is now the general practice from one end of Scotland to the other.' The farm, once a 'very wild and uncultivated piece of land', had a 'degree of neatness and garden-like culture, which in farming had not been seen before. Persons of every description came from every quarter to gratify their intellectual curiosity, as well as to get information.' This description was given after Hutton's death by a writer and advocate called Robert Forsyth, as a prelude to a harsh attack on Hutton's geological theories.

Inspired by the success of his modern farming methods, Hutton began his own scientific studies of agriculture. To investigate the effects of light and heat on growing plants, he tried such experiments as growing radishes and carrots in the dark. He explored the effects of climate and soil fertility on the quality and quantity of his seeds. He experimented with salt, seaweed, dung and even coal ash as fertilizers. He discussed these studies with friends in Edinburgh, and would visit the farms of other Improvers, sometimes in the company of his friend Dr James Russel, who would become another lifelong companion.

Hutton also began a detailed series of studies of the weather, concerned with how climate, especially temperature, affected plant growth. This became a lifelong project. Even back in East Anglia he had taken the temperature of springs along the coast, trying to establish a relationship between latitude and temperature. Now, at Slighhouses, he could be seen wetting thermometer bulbs and holding them up to the east wind, to see how evaporation caused a drop in temperature.

Hutton's experiences as a farmer greatly deepened and enriched his understanding of the natural world. Harking back to his medical studies, when he had been struck by the analogies between the human body and the body of the Earth, he tried to study the way dynamic forces – notably light, heat and the chemistry of the soil – operated together to support life. He was becoming fascinated by what we would today call *holistic* properties: how the world as a whole operated. He came to feel that most naturalists paid too much attention to classification and analysis, as if they were trying to break down the natural world into fragments.

Such reflections, laid over the strata of experience and philosophy in Hutton's still young mind, were already consolidating into a startling new geological insight – an insight that would shape the rest of his life, and ultimately the future of geology itself.

In the end, it came out of a crisis of faith.

9

'The Work of Infinite Power and Wisdom'

In the solitude of his farmhouse, Hutton brooded over the nature of God.

The doubts planted by Maclaurin's Deist lectures had deepened. The simple Christian beliefs with which he had grown up had long dissipated: 'There is nothing of the Christian left about me except some practice of prayer and piety.' But he seemed to have been searching for a new certainty: 'Faith, faith of all things is what I want most,' he wrote. 'I ha'nt a single grain of it to do me any good.'

By now Maclaurin's teaching of the essential beneficence of God and the perfection of His natural law seemed to accord more deeply than traditional teachings with Hutton's experience of the real world. After all, according to contemporary readings of scripture, the Earth was nothing but a ruin, a relic of a better time: Hutton would write, 'Philosophers observing an apparent disorder and confusion in the solid parts of this globe, have been led to conclude, that there formerly existed a more regular and uniform state, in the constitution of this Earth; that there had happened some destructive change; and that the original structure of the Earth had been broken and disturbed by some violent operation, whether natural, or from a supernatural cause.' Thus Burnet's Earth had been created perfect, but had, steadily or spectacularly, decayed ever since: even such beautiful features as Alpine mountains were just the ruins of an original Edenic landscape.

But Hutton himself had coaxed crops from the ground with his own fingers; he himself had turned an unpromising piece of hillside into a productive farm. Earth, it seemed to him, was not a wreck, but was generous and bountiful.

Besides, why would God purposely create a world that instantly fell into ruin? Did Genesis not say of the creation,

'And God saw that it was good'? What was *good* about a wreck? Wasn't it more likely that God would make a world that was in fact *perfect*, and stayed that way? And if God made the world, what did He make it *for*?

Hutton gradually became convinced that he needed to understand the divine design of the Earth. In doing so he constructed a new and crucially important theory that would cement his own reputation and place in history, and has shaped geology ever since. And, it seems, he found new faith.

Hutton would not publish his theory until the 1780s, but he had formulated his key ideas much earlier – possibly by 1760, when he was just thirty-four.

Playfair regretted the fact that even by the time of his biography, 1803, Hutton's papers 'do not afford so much information as might be wished for'. But Playfair refers to 'sketches' of an essay on 'The Natural History of the World' that were evidently written around 1760. Hutton's Edinburgh friend Joseph Black would, in 1787, send an abstract of Hutton's theory to an interested Russian princess, noting, 'Dr Hutton had found this system or the principal parts of it more than twenty years ago and he has found reason to be more and more confirmed in it by his study of [rocks] ever since that time.' So Black also puts the date of the theory's formulation at before 1767.

Hutton's 1760s sketch of his theory has not survived, but given Playfair's account, Hutton's own words of his first formal presentation of the theory, and what we know of Hutton's experience, I think we can reconstruct the pattern of his thinking at the time with reasonable reliability.

The Earth – like a human body – is a messy object. The evidence for divine design is much more easily visible in the sky above us, in the clean motions of the heavenly bodies. As Aristotle himself observed, 'Order and definiteness are much more plainly manifest in the celestial bodies than in our frame; while change and chance are characteristic of the perishable things of Earth.'

But if Earth has a purpose, what is it? Surely, Hutton decided at last, a perfect world-machine could have no higher purpose than to sustain life.

Fine – but if Earth is a life-support machine, at first glance it is a very imperfect one. Slowly, Hutton began to formulate what Stephen Jay Gould has called 'the paradox of the soil'.

For life, soil is everything, the substrate of existence. We couldn't live on a ball of rock: 'A solid body of land could not have answered the purpose of a habitable world,' Hutton would write in 1785, 'for a soil is necessary to the growth of plants.' That is why (according to Hutton's developing argument) erosion works on the surface of the world – to make soil out of rock. 'A soil is nothing but the materials collected from the destruction of the solid land ... Therefore, the surface of this land, inhabited by man, and covered with plants and animals, is *made by nature* to decay, in dissolving from that hard and compact state in which it is found below the soil' (my emphasis).

Erosion doesn't stop, however, when the soil is created. Because of erosion, as the Norfolk farmers had taught Hutton, the soil itself is washed away: the process that made the soil ultimately destroys it. Of course, there are more rocks to be weathered to sand and clay, more soil will be manufactured to replace what was lost. But erosion continues relentlessly. Given enough time, erosion will continue until 'the heights of our land are ... levelled with the shores [and] our fertile plains are formed from the ruins of the mountains.'

Eventually, if this goes on, the world as an abode of life will have an end-point: 'If the vegetable soil is thus constantly removed from the surface of the land, and if its place is thus to be supplied from the dissolution of the solid Earth, as here represented, we may perceive an end to this beautiful machine.' The very process that sustained life seemed doomed eventually to destroy it: *that* was the paradox of the soil.

But if this was so, Somebody had made an error: 'If no ... reproductive power, or reforming operation, after due enquiry, is to be found in the constitution of this world, we should have reason to conclude, that the system of this Earth has been intentionally made imperfect, or has not been the work of infinite power and wisdom.'

Hutton's key intuition was that perhaps the world did not suffer only decay: perhaps it also had the capability for *repair*.

Perhaps the world was like the fields he tended, sustained by his own careful crop rotation and fertilisation. Or perhaps it could even be compared to the Newtonian cosmos. Gravitation acts as a restorative force against the planet's inertia, the two effects together keeping the world on its life-sustaining orbit around the sun; there might similarly be some restoring force to balance the processes of geological decay.

Or perhaps, as Hutton had hinted in the title of his medical thesis, the Earth could be compared not to a worn-out machine, but to a human body. We are born, we develop, we age, we sink into decrepitude and we die: each life is a narrative of unique and unrepeated events. And, like our decaying bodies – according to one reading of scripture – Earth is proceeding from a state of perfection through decay into lifeless ruin: the Earth's meagre six thousand years are like humanity's three score years and ten.

But our bodies also exemplify another model of time: the old pagan idea of a history of unending cycles. Our hearts beat, our lungs fill and empty, our organs recover from injury and sickness, the blood in our veins circulates. Was it possible, then, that something of this could be reflected in the beneficent Earth?

So, drawing on his varied background and starting from first principles – that the final cause of the Earth is to sustain life – Hutton *deduced* that it must have some mechanism of repair from erosion, just as Harvey had once deduced the existence of capillaries in the body, then undetected, to complete his model of blood's circulation. The task now was to find that mechanism: 'This is the view in which we are now to examine the globe; to see if there be, in the constitution of this world, a reproductive operation, by which a ruined constitution may be again repaired, and a duration or stability thus procured to the machine, considered as a world sustaining plants and animals.'

In his geological observations of sedimentary rocks, Hutton had glimpsed just such a mechanism. He had compared the gravel and sand produced by erosion with the visible components of sedimentary rocks. They looked identical (save that the rocks had been cemented together by some mechanism he had yet to understand). Sedimentary rocks, then,

were formed from the erosion of some previous landscape,
whose rubble had been compressed and made into new rock.
And this must have happened under the sea. 'We find the
marks of marine animals in the most solid parts of the Earth,
consequently, those solid parts have been formed after the
ocean was inhabited by those animals, which are proper to
that fluid medium.'

That couldn't be the whole story, however. If the land was
washed away, even if the debris was consolidated into rock on
the sea bed, the Earth would soon be worn flat: 'We would thus
have a spheroid of water, with granite rocks and islands
scattered here and there.' Somehow the new rocks formed
from the destruction of the old must be raised *above* the
surface of the water. It had to be this way, if Earth was to stay
habitable – and perhaps Hutton, exploring the twisted, ancient
rocks of the Scottish Highlands, had wondered about the
mighty forces that could have produced such distorted
elevations.

This was Hutton's picture, then: rocks decayed through
erosion, the rubble was consolidated into new rocks, and then
somehow uplifted to make new lands – erosion, deposition,
consolidation, uplift. And cupped in the heart of this immense
rocky machine, the priceless soil that sustained life was subtly
created.

Thus, arguing from final causes, Hutton had resolved the
paradox of the soil, and arrived at a startlingly new view of the
Earth and its operations. His nascent theory of the Earth was a
synthesis of all his experience – the Deistic teachings of
Maclaurin, his microcosmic analogies between the planets and
the Earth and the human body, the benevolence he saw in
nature as a farmer. His design arguments, so alien to modern
thinking, had proved essential, for they had provided the
framework that enabled him to consolidate his puzzling
observations and construct his hypothesis.

Perhaps he had been hungry to find some such vision.
Throughout his life he had been inquisitive, scientifically
restless. He had, after all, studied under the great Maclaurin,
and at Europe's leading centre for medicine; he must have felt
as if his brain was rotting away, stuck out here on the farm.
Now, in the yawning chasm between the reality of Earth as he

had experienced it and the accepted theories, he felt he might have found a great intellectual contribution of his own, a unique vision. And stranded in a very imperfect world – a world in which his schooling was disrupted by the posturing of a prince, and in which his complex personal problems forced him to grub at the ground to make a living – perhaps a deep longing was planted in Hutton's mind to find an order in the mechanisms of the Earth comparable to that which Newton has found in the heavens.

Whatever the motives, it was a beautiful vision. With his 'series of great natural revolutions in the conditions of the Earth's surface', as Playfair described it, Hutton had made the world the mirror of the Newtonian sky. But, despite our retrospective framing of Hutton as if he was a man of our own time, it was *not* a modern scientific hypothesis; it was hardly possible that it could have been. Hutton's theory of the Earth was an argument about the nature of God. And Hutton must have known that a vision, however beautiful, wasn't enough.

Hutton was a good and successful farmer, but never a rich one. By the mid-1760s, his farms were returning a profit, but only a modest one.

Hutton's situation now changed. He had been elected to a committee which was to supervise the construction of the Forth and Clyde canal. He had long taken an interest in the heated debates over this project, and he must have been pleased to have found an avenue to exploit his geological understanding – and an excuse to get involved in the affairs of the city again. His financial circumstances also began to improve. The sal ammoniac business had at last, under Davie's careful stewardship, become seriously profitable.

By about 1765, this business, and perhaps other ventures, had freed Hutton from the need to earn his living by farming, and he began to plan a move back to the city. Another motive for this was his dissatisfaction with his Norfolk ploughman, whom he hoped to train up as a farm manager: 'As he could neither manage the business of my farm, nor act properly under the management of another, I gave up farming,' he would write. Meanwhile, after thirteen years, perhaps the

indiscretion that initially drove Hutton out of Edinburgh was
buried sufficiently deeply in the past for him to return without
stigmatisation.

At last, in 1767, aged forty-one, Hutton moved back to
Edinburgh. He kept the farms, though, running them under a
series of tenants or managers, and maintained an interest in
their upkeep.

Hutton's years as a farmer had made him. He had been
driven to his farms reluctantly, as a confused and unhappy
young man, but he had made the best of what he had been
given. With pioneering techniques, he had greatly improved the
condition of his land, and his agricultural studies had become
the focal point of his interlocking interests in chemistry,
meteorology, geology and botany. His achievements showed
the quality of his character and his mind.

Hutton never forgot what he had learned on his farm. In
1774 he would write from Bridgnorth to his friend George
Clerk-Maxwell in Edinburgh, reflecting on the agriculture of
the Welsh borders: 'Lord what an inclosed grass country I have
seen: you would not think that corn had been so dear for some
years by the aspect ... no wonder, they are bad at that trade
here, I have seen, more than once, five great waggon horses
yoked endwise like a string of wild geese, – doing what? What
I could not have believed if I had not seen drawing two
harrows, but for what purpose God knows.'

The most important legacy of his farming years, however,
was undoubtedly his geological intuition. As Einstein would
say of relativity, Hutton must have felt it was a theory too
beautiful to be false. But nobody knew about it yet – and
Hutton knew enough about the world of ideas to understand
that if he wanted to join the growing company of respected
thinkers in Enlightenment Edinburgh, if he didn't want to be
dismissed as another spinner of dreams like Burnet, he would
have to establish his theory on firm foundations.

In Hutton's time, however, the organised body of knowledge
we know as modern science had barely begun to be assembled.
Atomic theory, for instance, would not be generally accepted
for another century – in fact, as late as 1870 there were still
attempts to formulate a theory of chemistry without atoms
at all. Nobody understood the nature of heat; it would be a

century before modern thermodynamics was founded. In an intellectual climate still dominated by the notion that the Bible was the ultimate source of all wisdom, it wasn't even clear *how* to think about the world. Today it has become a cliché to say of some new scientific discovery – water under the deserts of Mars, or a new wrinkle in the human genome – that the scientists will be forced to 'rewrite the textbooks'. How much harder it was for Hutton in a time when the textbooks had yet to be written!

As he looked ahead to his new life in Edinburgh's energetic bustle, Hutton must have begun to understand that in order to establish his geological theory, he would have to become an expert on almost everything else as well.

TWO

Consolidation

10

'Assemblies of good fellows'

In December 1767, Hutton returned to Edinburgh to live with his three sisters. From now on he would live permanently in the city of his birth, and would build a home for himself and his sisters in a new development called St John's Hill, facing the spectacular geology of Salisbury Crags.

Free of the need to scratch for a living, Hutton took up the enviable life of an Enlightenment polymath. He was forty-one years old.

It had taken a generation, but the Union had at last transformed Scotland. Though nationalist dreams would linger, and for all the cross-border friction that lingers to this day, the Scots found themselves living under a strong, competent and stable government, and one that kept itself sufficiently remote, so long as there were no such unpleasantnesses as Jacobite risings, to allow the Scots to run their affairs largely as they liked.

As a result the economy was expanding fast. Scottish merchants had entered the Atlantic trade with the English colonies in America, a lucrative opportunity which had always been closed to them before the Union. A new range of goods, including tobacco, molasses, sugar and cotton, flowed into the country, and new exports, including finished goods like linen and cotton products, began to flood out. The 'Tobacco Lords' of Glasgow would dominate the market in their commodity, and become hugely wealthy in the process.

Now the city of Edinburgh was expanding too. By the time of Hutton's return, the population of the city had doubled since the Union with England, but they were all jammed into the same old medieval space. It was both 'the most picturesque (at a distance) and nastiest (when near) of all capital cities,' said Thomas Gray. The Royal Mile, the spine of

Castle Hill, was perpetually thronged by people, animals, vehicles and garbage. Away from the Mile you entered a labyrinth of shadowy, twisting streets of blackened houses and tenements, with the lower classes and servants jammed into the upper and lower storeys, and the middle- and upper-class folk, including nobles, in the middle levels. You were liable to trip over pigs and sheep in the street, and if you heard the cry of 'Gardy loo!' from an upper window you ducked or ran, for a chamber pot was about to be emptied into the road.

Things were changing, however. In these peaceful times there was no longer a need for the city to be restricted to its old defensive position around the ridge of Castle Hill. So in 1766 the planning of a new development, on the hundred acres of land above the North Loch, had begun.

Unlike the chaotic and disorderly old city, the New Town would be rigidly organised in the modern style, on a rectangular grid based on three long and wide avenues. The political agenda was clear from the naming of the streets: the three great avenues were to be George Street, Queen Street and Princes Street (the latter named after the Prince of Wales and his brother). Nevertheless the new development was uniquely Scottish. Unlike similar developments in England or France, the planning of the New Town left no room for huge aristocratic estates. Only one prominent noble moved in; the rest of the development was occupied by Edinburgh's middle and commercial classes.

The final phase of new Edinburgh's development was the beautiful Charlotte Square in the westernmost sector, an inspiration of Robert Adam. The young man who had once helped Colin Maclaurin plan the defence of the city against the Bonnie Prince was now a leading architect. Adam's clean and civilising designs, inspired by the ruins of antiquity, would make him the most famous and important architect in Britain.

It was against the background of the elegant new city, and in the heads of a literate, independent-minded and newly prosperous populace, that the Scottish Enlightenment would bloom, causing Voltaire to say, 'It is to Scotland that we must look for our idea of civilisation.' And it was in this sparkling Enlightenment Edinburgh that Hutton now immersed himself.

Even before his arrival, Hutton was well known in Edinburgh circles. His old friend John Clerk of Eldin was there, along with his brother George Clerk-Maxwell, with whom Hutton had been working on the project to drive a canal between the Firths of Forth and Clyde. There had been something of a 'canal mania' in Britain in recent years, as the use of a canal could cut the costs of transporting coal by half. Later Hutton and Clerk-Maxwell became shareholders in the Forth and Clyde Navigation Company. The canal's construction was begun in 1768, and Hutton would continue on the project until 1777. This overlapped with his other interests, for instance in seeking raw materials for underwater-setting cement.

Another of Hutton's circle in Edinburgh was James Lind, the doctor who had discovered the treatment for scurvy. Lind had planned to sail on Captain Cook's second voyage, but withdrew and went to Iceland instead. Later he became physician to the royal household at Windsor. James Russel, Hutton's old friend from his farming days, was here too. Russel, a surgeon-apothecary, became a professor of natural philosophy. Hutton also became friendly with Russel's son. The younger Russel became a surgeon, eventually rising to the position of President of the College of Surgeons in 1796, but he always remained Hutton's family doctor.

Hutton was quick to join the Philosophical Society, which had been established in 1737 by Colin Maclaurin, Hutton's Newtonian mentor. Maclaurin had proposed that the Society for the Improvement of Medical Knowledge, set up in 1731 by the anatomist Alexander Monro, should be reformulated to become The Edinburgh Society for Improving Arts and Sciences and Particularly Natural Knowledge – or, as it became more popularly known, the Philosophical Society. The untimely death of Maclaurin cost the Society some of its momentum. Nonetheless its papers included one read by Boswell for Dr George Young, once Hutton's medical tutor, on 'bones found in the ovarium of a woman', and a letter from Dr B. Franklin to D. Hume, Esq., on 'the Method of Securing Houses from the Effects of Lightning'.

Hutton read many papers to the Society, of which only two became more widely known, on botany and artillery. But it seems likely that he speculated widely on philosophy, epistemology, language, agriculture, geology, chemistry and other subjects which would show up in his later writings. In 1769 Hutton was involved in the Society's attempt to organise an expedition to observe the transit of Venus across the sun, but the project turned into something of a fiasco (the transit was observed successfully by Captain Cook).

For Hutton, the most significant paper of all concerned 'Experiments upon Magnesia Alba, Quicklime, and Some Other Alkaline Substances', submitted in 1755 by one Joseph Black MD.

Black, whom Hutton may first have met as a cousin of Russel's, was born in Bordeaux, two years after Hutton. His father, native of Belfast but of Scottish descent, was a wine merchant. Black trained in medicine and science at Belfast and Glasgow; by 1756, he had taken posts in chemistry and medicine at the University of Glasgow.

Unlike Hutton, Black also practised his medicine. In 1771 Black would attend Walter Scott's birth: the nurse had concealed the fact that she had consumption, but she confided in Black, and the doctor's expertise saved both nurse and child. Black's kinsman Adam Ferguson would write, 'Without flattery, or uncommon pretension to skill, [Black] won the confidence of his patients, and, with unaffected concern for their benefit, was often successful in mitigating their sufferings, if not in removing their complaints.' You probably couldn't ask more of a doctor in the late eighteenth century.

Black's achievements as a chemist, though, are his most notable. His careful and precise quantitative work anticipated the methods of modern chemistry. Through experiments on heating magnesium carbonate, quicklime and other substances, he proved the existence of carbon dioxide.

What would prove still more significant were Black's notions on latent heat. While still a student, Black had noticed that if you warm up a block of ice it melts without changing temperature. This meant, Black argued, that heat must have flowed into the ice and combined with its particles, becoming 'latent' – hidden. Black's work on this hidden heat would be one of the foundation stones of the modern science of thermodynamics, which has led to a proper understanding of the working of heat engines of all kinds.

For his part, Hutton was electrified by Black's quicklime experiments. Hutton's theory needed a way for rocks to be transformed from one form to another, and – though this hadn't been their intention – Black's experiments had shown that this could very well happen under the influence of heat, opening up a new avenue of thinking for Hutton.

Mutual interests in chemistry and geology drew Hutton and Black together: Black's chemistry lectures referred to plenty of geological concepts, including the formation of strata and mountains, and experiments on quartz, flint and basalt. Black's ideas prove profoundly significant in shaping Hutton's own thinking – the pair discussed Hutton's theories of the Earth over many years – and eventually Black 'subscribed entirely to the system of his friend.'

Black became perhaps Hutton's closest friend. Like Hutton, he was a bachelor, but their personalities contrasted strongly. Playfair noted that 'Black was serious, but not morose; Hutton playful but not petulant.' Black, the systematic and quantitative chemist, was careful and cautious in his speculations, while Hutton 'could be in the air [and] speculate beyond the laws of nature'. A caricature by John Kay, made when they were both in their sixties, shows a periwigged Black listening to Hutton – arms folded, without a formal wig, his bald head fringed by unruly hair. They are both smiling, and their relaxed friendship is evident.

While the physical scientists like Hutton would extend Newton's legacy in their thinking about the natural world, much of the moral flavour of the Enlightenment was provided by thinkers like Adam Smith and David Hume, who laid the foundations of what today are known as the social sciences: sociology, history, psychology, economics.

Hutton soon became a close friend of Adam Smith. They may have met as early as 1751, when Smith was giving public lectures in Edinburgh. Smith had been born on Kirkcaldy in 1723. His father had been one of that very unpopular breed, the new post-Union customs inspectors. Smith's *Wealth of Nations* established him as the first great modern economist. He endeavoured to apply a Newtonian analysis to people; starting from basic principles of humanity and human nature, he sought to derive principles of law, economy and international relations. Young Smith's observations of how smugglers found ever more ingenious ways to evade the government's strictures were no doubt an early lesson for him in the power of self-interest in motivating human beings.

The philosopher David Hume, some fifteen years older than Hutton and twelve years older than Smith, was the son of a modestly propertied laird. His mother encouraged him to go to university to study law, but his omnivorous curiosity soon drove him to give this up, and he turned to 'books of reasoning and philosophy, and to poetry and the polite authors'.

In 1734 Hume began a three-year stay in France, mostly at La Fleche where Descartes had studied a century earlier. There, by the age of twenty-seven, Hume completed his first major work – a three-volume *Treatise on Human Nature* – which set out the philosophical themes that would dominate his life's work. But the treatise, he said, 'fell dead-born from the press'. Over the next decade he expanded and rewrote his ideas, but without the sustenance of an academic position Hume had to find other work for a living. He served as private secretary to a French general who was planning an invasion of Canada, as emissary to the courts of Vienna and Turin, and as a librarian. His writing on various subjects gradually won him a following, most notably for his six-volume *History of England*, published by 1762.

It is not clear whether Hutton ever met Hume. There were certainly links between them: John Clerk of Eldin was the brother-in-law of Robert Adam, who in turn was a close friend of Hume. Hutton and Hume could have met in 1766 during Hutton's farming years, when Hume made a short visit to the family farm at Ninewells, only a few kilometres from Slighhouses.

Whether they met in person or not, Hume – and particularly his thinking about the nature of knowledge – had a profound influence on Hutton.

Hume saw philosophy as being an experimental science of human nature, and in progressing that science he drew inspiration from Newton as well as from such predecessors as Locke and Berkeley. We can only know anything, after all, through the operations of the mind, and so to understand the limits of our knowledge we have to begin with an awareness of ourselves as observers: 'As the science of man is the only solid foundation for the other sciences, so the only solid foundation we can give to this science itself must be based on experience and observation.'

Hume separated the contents of the mind into 'impressions' and 'ideas'. Impressions can rise from external sensations or from internal cognition – an unbidden memory, say. They include 'all pure sensations, passions and emotions, as they make their first appearance in the soul'. Ideas, on the other hand, are the products of analysis and classification. If impressions are the input to thought, ideas are the output. We need to aid perception and action: being able to formulate the idea of a lion from a jumble of impressions is obviously an aid to avoiding becoming its lunch. To some extent, though, ideas are disconnected from impressions, and therefore from reality.

This has implications for causality. When we say 'A causes B' we are really talking about a linking of ideas, not of impressions, still less of objects out there in the real world. Hume firmly believed that events themselves are causally related, and that they will be related in the same way in the future as the past – so if a falling glass broke yesterday, an identical glass will break today. These 'natural beliefs', he said, are shared by everybody, and are in fact necessary to survive. But he pointed out that we actually have no evidence for such beliefs, and that there can never be such evidence. He wasn't denying belief, but questioning unjustified certainty, and he was trying to make philosophers and scientists think hard about what they regarded as truth.

Hume's work was very important for Hutton – who was dealing, in his geology, with remote realms like the depths of the Earth, or distant ages past, of which we can have no direct knowledge. How, then, is it possible to know anything about them at all? He would find tentative answers in Hume's thinking.

More controversially, Hume also asked very hard questions about God. In his *Dialogues Concerning Natural Religion*, he asked why we should look for 'mind' as the organising principle of the universe. Maybe the order we perceive just *emerged*. 'For aught we know a *priori*, matter may contain the source, or spring, of order originating within itself, as well as the mind does.' This sentence was written a full century before Darwin came up with a way to show how the organised complexity of life could indeed emerge from 'matter itself', without the need for any guiding mind.

Hume could at times be very harsh about institutional religion: 'Survey most nations and most ages. Examine the religious principles which have in fact prevailed in the world. You will scarcely be persuaded they are anything but sick men's dreams.' No wonder the popular take on Hume's work was that he was denying God. This led to him being refused university posts. Hume and Hutton, in fact, were the most significant figures of the Scottish Enlightenment *not* to hold formal academic positions.

Once Hume took a short cut home to the New Town across the sticky bog left by the draining of the North Loch. He fell into the mud and was unable to extricate himself. He asked a passing fishwife for help, but she recognised 'David Hume the atheist', and wouldn't help him out until he had repeated the Lord's Prayer and the Creed of the Apostles. Hume later took great pleasure in telling the story against himself. He refrained from publishing the *Dialogues*, however, until after his death.

There was, of course, more to the Scottish Enlightenment than high thinking and public good deeds.

Edinburgh, like other urban centres, was dominated by the gentlemen's club: an institution Samuel Johnson defined as 'an assembly of good fellows, meeting under certain conditions'. Some clubs had dedicated premises, but mostly their meetings were held in coffee bars and hotels – or, more typically, taverns. Many of the clubs were dominated by men like Hutton, of the new middle classes, but some focused membership on the aristocracy, landed gentry or military officers. Clubs evolved their own rules and rituals, acted out with solemnity, mock or otherwise. New recruits were often subject to initiation rites. Many of the clubs invented cod histories, and hid their activities from the general gaze, thus generating an aura of exclusivity and mystery.

The members of a club might have some shared purpose. Hutton would no doubt have been aware of the Circulation Club, whose members commemorated William Harvey's discovery at annual dinners on his birthday with a circulation of glasses containing other vital fluids. In about 1770, Black formed the Poker Club, containing judges, lawyers and soldiers as well as scholars. This little group – named after the humble tool that stirs things up in the fireplace – advocated the notion of an armed militia, based on 'a conviction that there could be no lasting security for the freedom and independence of these islands, but in the valour and patriotism of an armed people'. Black's generation remembered too well the helplessness of the Edinburgh volunteers before the uncouth forces of Prince Charlie's army.

Male conviviality, though, was the clubs' primary *raison d'être*. Almost all excluded women, and certainly children: a gentlemen's club was a place to escape the responsibilities and constraints of the household – to talk dirty, act badly and, especially, to get drunk.

Enlightenment Edinburgh was awash. The drink of preference was claret, a legacy of Scotland's historical ties to France. Even after 1707, when the English fashion for port and sherry began to move into Scotland, sticking to claret came to seem a patriotic act – whisky was still seen as a crude local brew.

The quantities consumed could be heroic. A gentleman would be labelled a 'two- or three-bottle man', depending on his consumption over dinner. Then after the meal, the protocol was for the host to toast each guest in turn – and then each guest toasted his host, and each of the other guests. A little mathematics shows that even a modest gathering of, say, five people would require a total of twenty toasts to be raised. But the booze wasn't restricted to dinner. Drinking, according to one account, 'engrossed the leisure hours of all professional men, scarcely excepting even the most stern and dignified'. People would drink over business deals and legal matters, even preparing for a day's work on the bench.

Not all the clubs were particularly intellectually minded. The notorious Beggar's Benison had been founded out in the sticks in Fife, where bored local merchants and landowners toyed with bawdiness, irreverence, and a little playful Jacobitism. Its title celebrated the exploits of the Stuart King James V. Once, the legend went, James came to a burn he was reluctant to cross, and paid a beggar woman to carry him over. She thanked him with her benison, or blessing: 'May prick nor purse never fail you.' The club would become infamous for the obscenity of some of its activities, not least the hiring of 'posture girls' for the edification of its gentlemen. By 1752, the club had a branch in Edinburgh. One historian would claim that 'it is difficult to say who, of any prominence in literature or society, at that time, was not a member of the Beggar's Benison.' The Benison even granted an honorary membership to the Prince of Wales – later George IV – a notorious socialite who gorged, drank and flaunted his mistresses in public.

It is not known whether Hutton was a member of the Benison, but as far back as 1741 his good friend Sir John Clerk had been a correspondent of the Benison's 'sovereign' over a Roman phallus that had turned up in Scotland (Clerk was a noted antiquarian). Hutton might have enjoyed the Benison's cheerful adoption of Ussher's timescale for the invented origins of their society; as sex was created with Adam and Eve, so was the Benison's 'most ancient and puissant order'.

Of course, there was a downside to all this. In the 1780s William Creech, eventually Hutton's publisher, would satirically portray a dark mirror of the gentlemen's clubs – the 'Jezebel Club', a society for the prostitutes whose numbers had multiplied during the years of the Enlightenment. The Jezebel Club had ongoing vacancies, Creech said harshly, for many of its members were dying, 'decayed', before they were twenty.

In this milieu Hutton lived his life to the full.

Freed of the strictures of husbandry, with no pressure to earn money, and with a bachelor's lack of domestic constraints, Hutton evolved a comfortable lifestyle. He would rise late. He was endlessly busy with his intellectual, public and business projects. He wrote continually – he 'was in the habit of using his pen continually as an instrument of thought' – and he left behind an 'incredible quantity' of manuscript, much of it never intended for publication.

In these progressive times there was a good deal of public spirit about. Hutton showed no signs of ardent nationalism, but he was clearly proud of the rebirth of Scotland, and he advocated projects like the canal scheme which he believed would benefit his country. In 1777, he published a short work called *Considerations on the Nature, Quality, and Distinctions of Coal and Culm*, his first publication since his medical thesis. Parliament had laid down different rates of sea-transport tax for coal-dust, depending on whether it was good enough for 'coal', used for domestic hearths, or only good enough for 'culm', fit for kilns. Since there was no test to distinguish these abstruse qualities, Scotland found itself penalised. In his pamphlet Hutton described a simple test based on nothing more than throwing a handful of dust onto a red-hot shovel and seeing what happened. This quickly resolved the situation and brought financial relief to the industry.

Meanwhile, Hutton kept up his interests from his farming days. He became interested in the political aspects of agriculture. Despite his friendship with Adam Smith, he believed in government intervention in agriculture, as it was too important to be left to market forces and chance: 'The husbandman maintains the nation in all its ease, its affluence and its splendour,' he wrote. But farmers too had a responsibility for the public good. Rotation of crops, ensuring equal acreages of different crops at any given time, would help keep prices stable.

The urge to experiment did not leave him. Visitors to his Edinburgh garden would see patches of cauliflower, half of which had been grown in unfertilised soil and half in soil fertilised with nitre. He was often consulted on agricultural issues, and he helped his friend Joseph Black with experiments on unusual types of soils, and on dyes made from vegetable products.

He kept up his climate studies too. His new house was only a few hundred metres from Arthur's Seat, and he took to walking up the great volcanic plug in all seasons and all weathers, to study the relationship between altitude and temperature. These studies, involving tediously obtained records compiled over long periods, offer another glimpse of Hutton the working scientist: his geological theories may have been bold and his manner impulsive and speculative, but he was capable of persistent, careful and thoroughly practical experimental and observational work.

The evenings were the focus of Hutton's life. He was evidently popular in the Edinburgh clubs: he lacked self-consciousness, which would 'sometimes lead him into little eccentricities, that formed an amusing contrast with the graver habits of a philosophic life'. He took great pleasure in 'domestic society'. Hutton's circles of friends included 'accomplished individuals of both sexes' – in fact, in his moral thinking, he had come to believe in 'the importance of the female character to society, in a state of high civilisation'. He had evidently got over the unhappy misogyny that had afflicted him after his disastrous love affair, and the loneliness of his farming days was a memory.

Playfair records that his conversation was always lively, forceful and informed, and full of wit and speculation. Judging from his few surviving letters (he was not a great correspondent) his speech would have been rambling, entertaining, and taken at a rush. For example, in a letter of 1774 to George Clerk-Maxwell he described a visit to Bath in a mixture of jokes, bawdiness, mild blasphemy, with a little geology thrown in: 'To be sure Bath is the most curious & valuable spot in Britain without exception – for a place to have had the chaude pisse [the hot springs] for many thousand years and be as embonpoint as those that have shitten snowballs ever since the immaculate conception of the blessed virgin tho a little paradoxical is notwithstanding very true – Some places have lime to a vice like the chalk downs, but here tho you see nothing but lime there is an abundance of fine strong soil ...'

Despite such evidence Playfair paints a portrait of a restrained, almost ascetic character. Hutton dined early and usually at home, Playfair would have us believe, and 'he ate sparingly, and drank no wine'. Robert Louis Stevenson, born 1850, would look on Raeburn's portrait and remark on 'Hutton the geologist, in quakerish raiment, looking altogether trim and narrow, as if he cared more about fossils than young ladies'.

However, Playfair knew Hutton only in the geologist's middle and old age, and, writing after Hutton's death and in a moral and political climate transformed by the horrors of the French Revolution, he was trying to present a sanitised Hutton to the world, a model Enlightenment gentleman whose scientific theories thereby carried a little more authority. Hutton's letters show us that behind Playfair's frozen image there was a human being – warm, impulsive, crude, funny, lustful, and frequently drunk. James Hutton was surely no libertine, but there is no reason to suppose he lived monkishly. I find it pleasing to believe that Hutton, who was working out his own essentially optimistic and uplifting vision of the Earth as a work of God's bounty, would have accepted the gift of his carnality in the same sense.

And above all, there was the geology.

Hutton continued to amass geological samples. The rocks were his 'treasures' and his 'ewe lambs'. As always, friends and acquaintances sent him material from Britain and abroad: he had rocks from Labrador, Sweden, Spain, Africa, the Urals, Gibraltar. Sir Joseph Banks, the naturalist who had travelled around the world with Cook, sent him material from Iceland. It was a relief for Hutton to move into his new house on St John's Hill; his collection was taking up too much space. Before that he had been 'obliged to make one chamber serve me for laboratory, library and repository for self and minerals, of which I am grown so avaricious, my friends allege I shall soon gather as many stones as will build me a house'. Hutton would examine his specimens chemically and under the microscope, and then varnish them to keep them looking bright. He would prune his hoard to focus on the most intellectually valuable samples. He would say, 'My ambition is to make a spacious library, where the books shall consist, in the ancient manner, of tablets of stone, and (without any mystical sense) wrote by the finger of God alone.'

His great theory of the Earth seems never to have been far from his mind. He had found no reason to question his basic conclusion, that Earth is a self-renewing machine whose purpose is to sustain life. But he still needed a way to bake rubble into rock, and an engine to drive his mighty continental uplift.

He found the answers, at last, in the machinery of the Industrial Revolution.

I I

'The power of heat is unlimited'

James Watt's first large-scale steam engine was erected in the
Burn Pitt colliery at Kinneil in 1765. Hutton visited Kinneil,
and as noted would later comment on marine fossils he
observed there.

Hutton had become a close friend of Watt. They probably
met through their mutual friend Joseph Black, or possibly
through the canal scheme; Watt spent eight years working as a
land surveyor on the project, marking out routes across
Scotland. Playfair tells us that Hutton 'had an uncommon
facility in comprehending mechanical contrivances ... He would
rejoice over Watt's improvements to the steam engine ... with
all the warmth of a man who was to share in the honour and
profit about to accrue to them.'

Watching the pumping pistons of Watt's prototype, Hutton
was mightily impressed with what he saw.

Hutton had come back to Edinburgh with two key questions.
He had in his head his vision of the Earth's dynamic cycles,
serving the divine purpose of making the surface habitable.
Very well. But if rocks were consolidated from the debris
produced by erosion – well, how? What agency could possibly
transform sand into sandstone, sea creatures into chalk, rubble
into conglomerate rock? And if the new rocks were uplifted to
make new lands – again, how? What could provide the tremendous energies necessary to erect mountains and lift continents?
There was no answer to be had in the conventional wisdom
that prevailed in Britain and across the Continent.

The still unnamed science of geology was slowly maturing.
As Hutton seems to have realised for himself, it had come to
be understood that the theories of cosmogonists like Burnet
would remain just imaginative fancies unless more information was gathered; and that could only come from a
painstaking examination of rocks in the field. Here geologists

in Europe took the lead, with careful studies of layers of rocks in Sweden and Germany, while Nicolas Desmarest explored the extinct volcanoes of central France.

Meanwhile, new theories of the Earth were devised, more soundly based on the new types of data and classifications of the rocks. Some of these derived strictly from scripture, others less so. But they all generally clung to the notion that Earth's rocks had all been deposited from some primeval fluid: even now, they were essentially descendants of the story of Noah's Flood.

Some of these theories, influenced by Hutton's old contact Rouelle in France, purported to explain the order in which rocks were found in the ground. Most of them held that the rocks had been laid down in a once-and-for-all fashion – and granite, the bedrock of the continents and mountains, was usually taken as the most 'primitive', the oldest rock of all. This, of course, ran counter to Hutton's own thinking; according to his slowly coalescing model, every rock type could be renewed, and nothing was truly 'primitive', unchanged since the formation of Earth.

There were many other problems with these soggy theories. What could the primordial fluid, in which all the rocks were supposed to have been dissolved, possibly have been made of? Certainly not water: 'If it is by means of water that those interstices have been filled with those materials,' Hutton would write, 'water must be ... an universal solvent, or cause of fluidity, and we must change entirely our opinion of water in relation to its chemical character.' In fact, Hutton knew there was *no* one solvent which was capable of dissolving all the minerals he saw in the rocks – nor even of dissolving all the different varieties you could find within one specimen, if you looked hard enough.

The compactness of the rocks presented a further difficulty. In conglomerate rocks – rocks composed of fragments of other rocks – there was no space between the component parts. So what had happened to the solvent fluid? Surely some of it should have been trapped inside the solidifying rocks. And why should the fragments of rock fit together so well? It looked more as if the rocks had *melted*, and then flowed, to fill up the spaces like toffee in a mould.

Black's 1755 experiment seems to have been the trigger that turned Hutton's attention to heat as the solution to these conundrums. Heat was certainly another way to turn rocks into liquid. When they had melted, the rocks would flow to fill all available spaces. Suppose, then, that the rubble of eroded rock and fossils was *baked* in some great subterranean oven to form limestone, chalk and the other rocks?

There were obvious objections to any heat theory. Consider coal, for instance. Here was a substance formed within the Earth, but which when brought to the surface is flammable. So how could it have been created by the action of heat? Wouldn't it simply have burned up under the ground?

Another problem was posed by limestone. In his 1755 experiment, Black, on heating a magnesium carbonate limestone, had succeeded in driving carbon dioxide out of the rock. Black was interested in the carbon dioxide he had thereby discovered, but what intrigued Hutton was the limestone itself. If limestone was formed by heat – and its crystalline structure certainly suggested that it had been – shouldn't its carbon dioxide have been driven off in the process? So how could it have still been there for Black to find?

An answer suggested itself in the work of Denis Papin, a French physicist and pioneer of steam-engine principles. While experimenting with pressure vessels – 'digesters' – Papin showed that water required more heat to boil when under pressure. So *pressure affects the action of heat.*

Perhaps limestone, then, was formed under great pressure – pressure that had trapped the carbon dioxide in its substance, even when great heat was applied to it. And perhaps, likewise, there was a way to bake dead wood into coal without burning it, if you again applied enough pressure – and surely Earth's interior must be subject to extremely high pressure?

To back up these tentative ideas, Hutton became interested in laboratory experiments on the effect of heat on rocks. Following his experiments on limestone, Black had already tried melting basalt. By 1772 Hutton was performing experiments of his own on zeolites, minerals in which water was trapped in cavities. Presumably in the kitchen of his Edinburgh home, he ground up the rocks, combined the dust with acid into a paste and boiled it to study its decomposition.

Hutton also asked Watt to use his steam-engine furnaces to heat a piece of iron to red heat and report whether it became brittle. Similar trials were proceeding elsewhere. Other experimenters melted lavas to form a variety of glasses, and the Frenchman Desmarest tried to melt specimens of granite.

As ever with these Enlightenment polymaths, some of the experimentation was motivated by profit. Watt, for example, had a share in the Delftfield Pottery Company in Glasgow. Porcelain is basically fine-grained earth mixed with glass. Many experimenters, inspired by research into Chinese manufacturing methods, had studied feldspar, a mineral common in granite, and kaolin – china clay – ingredients used by the Chinese. By 1768 Black and Watt were experimenting with these and similar substances, melting them in crucibles to see what resulted.

Meanwhile, it seems to have occurred to Hutton that subterranean heat might also be the solution to his other key problem: the source of the energy he needed if he was to raise continents. After all, he was a friend of James Watt himself: to see the potential of heat to do mechanical work, he only had to look around Watt's workshops.

The world's first steam engine was probably the *aeolipile*, invented in the first century by Hero of Alexandria, a Greek geometer and inventor. Hero set a metal sphere with two canted nozzles on an axis over a boiler. When steam spurted from the nozzles, the sphere spun around. It was a startling, anachronistic invention. But in the context of the times, the *aeolipile* remained no more than a toy, a potentially world-transforming idea soon to be lost in the Dark Ages.

The growth of industry in Britain in the seventeenth century created a strong motivation to find a new power source to replace water, wind and the work of human and animal muscles – a source preferably unaffected by geographical location and the weather. Without such a source, most factories were still limited to the rare locations suitable for running waterwheels.

The steam-engine principle seems to have been rediscovered by Papin (whose observations on the effects of pressure on boiling water had helped inspire Hutton's ideas on subterranean compression). Papin's experimental pressure cooker, his 'digester', was capable of creating a pressure difference which drew water up through a certain height: that is, the steam could be made to do work.

Thomas Newcomen's 1712 design was the first commercially successful steam engine. His steam pump was basically an upright cylinder fitted with a piston. When steam was allowed to expand into the cylinder it pushed the piston to the top of its stroke. Water was then squirted into the cylinder, cooling the whole apparatus down. The steam condensed and contracted, creating a partial vacuum inside the cylinder, which pulled the piston back down, ready for the next stroke.

Newcomen's engine was crude – it converted only one per cent of the steam's heat energy into mechanical energy – but it was a practical way of putting heat to work. Its basic design was unrivalled for fifty years. Even before James Watt was born, Newcomen's 'fire-engine' had become a workhorse in draining Welsh mines.

But it was Watt who would figure out how steam could be harnessed most effectively, thus changing the world.

Ten years younger than Hutton, Watt was born in Greenock. As a boy he had spent much time in his father's workshops, where he had his own tools, bench and forge, and made models of cranes and barrel organs – practical engineers always tell you that there is no substitute for the experience of cutting tin. At the age of seventeen, Watt aimed to become a maker of mathematical instruments. By 1757, he had opened a shop in Glasgow and was making compasses, scales and quadrants. He met Joseph Black, and began learning chemistry; he and Black would argue over the nature and applications of heat.

Before long, though, Watt branched out into making musical instruments. Perhaps the mathematical-instrument business didn't pay well enough. Soon he was making and selling flutes, violins, guitars, even a barrel organ, which still survives – and naturally, being a Scot, bagpipes. (It may be that Watt's morals were a little lax at this time. Among the bits and pieces in his workshop has been found a steel stamp for marking wooden goods with the maker's name – but the stamp bears not his name, but that of Thomas Lot, the Parisian doyen of flute-makers. A Lot flute would bring in a great deal more than a

Watt one. Was Watt passing off cheap imitations?

Watt's attention was drawn to steam engines by his friend John Robison. Robison, who studied under Black, was involved in another of the age's great enterprises; in 1762 he represented the Board of Longitude on the voyage to Jamaica which tested John Harrison's latest chronometer. Robison had wondered if steam engines could be applied 'to the moving of wheel-carriages, and other purposes'.

Intrigued, Watt took apart a model of a Newcomen steam engine, owned by the University of Glasgow. He laboured over the model for a year, trying to improve its performance. Black's notions of latent heat helped him understand the engine's operation. Watt eventually saw that the engine's inherent inefficiency came from raising and lowering the temperature of the whole working cylinder during the operating cycle from steam inlet to condensation. Watt took the steam out to a separate condenser and allowed it to cool and contract *away* from the main cylinder, making it unnecessary to heat up and cool the whole cylinder with each piston stroke – you only had to heat up the steam itself, while the bulk of the apparatus could be kept at steam temperature. The idea of the separate condenser was Watt's most fundamental contribution to steam technology. Even his first experimental engines, operating by 1765, cut fuel costs by three quarters.

Loans from Joseph Black had helped Watt build these first small-scale test engines. Now, through an introduction by Black, he entered a partnership with John Roebuck – the glassmaker who had provided Hutton and Davie with glass spheres for their factory – who offered to pay for development of the new engine if Watt could build a prototype for the foundry at Kinneil. Watt took out his first patent, on 'A New Invented Method of Lessening the Consumption of Steam and Fuel in Fire Engines', in 1769.

Watt's early years were blighted by a slow struggle for commercial success. Hutton was always kindly and concerned, advising Watt over such issues as how to approach Parliament to cover his work with patents: he wrote to Watt in 1774, 'I am sincerely of the opinion that a short account of the machine should be made out with the distinct estimate of its value compared with the common one and then represent that besides the invention of this uncommon advantage it has cost many years labour & expense to bring it to the perfection of utility.' And in 1788 Hutton monitored a steam engine designed by one William Symington which he believed breached Watt's patents.

Hutton also tried to find Watt lucrative positions in Scotland working for city councils, or even the police, and he advanced Watt's case with influential friends. Hutton became very critical of a system where appointments depended not on merit but on the support of those in power. Of one possible opening for Watt he wrote, 'I think it only needs to have a man properly bestir himself but that is what few political people do unless to serve themselves.'

Hutton was not above teasing Watt, though. In a note in 1774 he wrote, 'I write this to desire something to fill my vacuum which you know nature abhors – this is the reason that philosophers whose business it is to turn nature upside down have invented cylinders full of steam with condensers at their arse which is vexing whipping and spurring nature to work out of her ordinary course that these bougres [buggers] may sit idle on their arse ... Not so St Samson who was a holy man tho neither a philosopher nor a Bachelor god knows, he turned the mill himself and after lying all night with a whore at Garza he carried away the gates of the toon next morning on his back and all this without any subterfuge or second hand work.'

It was surely at Kinneil, if not earlier, that Hutton's lively mind, while pondering on the potential of Watt's crude but already powerful prototype, speculated that machines driven by heat could be mighty indeed.

The energies Hutton saw working in Watt's first machines clearly inspired him to think of the Earth as containing a tremendous heat engine. It was certainly an elegant solution to his dual problem, that the subterranean heat that created the new rocks in the first place should also be the agent that uplifted them: 'We may, perhaps, account for the elevation of land, by the same cause with that of the consolidation of strata.' And just as Watt's steam engines would soon empower the Industrial Revolution, so Hutton saw no end to the capacities of such immense energies. As he would say in 1785, 'The power of heat for the expansion of bodies, is, so far as we know, unlimited.'

But what *was* heat? Today we know that it is a phenomenon of the random motion of atoms and molecules. In a hot gas, molecules fly more rapidly; in a hot solid the molecules vibrate more enthusiastically about their average positions. But in Hutton's time, despite the pioneering work of Black and others, theorists struggled to explain the nature of heat.

After all, even the nature of matter wasn't understood. Some of Newton's contemporaries, particular Robert Hooke and Robert Boyle, were already 'atomists'. But others of Newton's followers had tried to devise systems of nature dominated by the motion of 'subtle fluids' or 'ethers'. These strange substances were supposed to permeate everything, and to underlie phenomena from gravity to electromagnetism and light. The interpretation of heat as a random motion of molecules (the modern view) became overshadowed by the notion of heat as a subtle fluid called *caloric*. This fluid would flow from cold places to hot: so when the sun warms your face, a gentle rain of solar caloric is actually falling on your skin. This wasn't a bad theory in its own right; the French scientist Sadi Carnot used it to derive his great theories in thermodynamics, including the 'Carnot cycle' description of the working of heat engines which is still used today. (In the end the caloric theory was beaten by careful and quantitative experiments on chemical combinations and heat engines by Thomson, Joule and others.)

Hutton absorbed all these ideas, and, following the pioneering work of Black, began to develop his own theoretical model of the nature and behaviour of heat.

Of course theorising, standing around admiring steam engines, and melting a few bits of rock in the lab, was all very well. But if rocks really were shaped by some subterranean source of heat, there ought to be more evidence out in the field. Thinking about this, Hutton's attention was drawn to basalt.

We know now that basalt is actually a common form of lava – that is, solidified magma, once-liquid rock. But in the standard diluvian theories it was held to have been laid down out of the primordial ocean, along with everything else. As many observers before Hutton had noted, basalt was usually found in *veins* that passed through layers of rock of other kinds. It often looked as if the basalt had been injected there in a liquid form, and then solidified in place. In 1764 Hutton had observed an example of this at Crieff, and in Edinburgh he found an example closer to home: in Salisbury Crags, in fact, just a few hundred metres from his front door. 'On the south side of Edinburgh I have seen, in little more than the space of a mile from east to west, nine or ten masses of whinstone [basalt] interjected among the strata.' Part of the exposure he found is now called 'Hutton's Section'.

Hutton became convinced that basalt had an igneous origin: that is, it must have been formed by the action of heat (the word 'igneous', like 'ignition', derives from the Latin word for 'fire'). If Hutton was to establish his heat argument he had to find instances of basalt and other rocks which showed unequivocal signs of the action of heat. And to show that the renewal of Earth was real he had to prove that 'primitive' rocks like granite, supposedly laid down before life had formed, weren't so primitive after all: he needed to find fossils in them. To do all that, he would have to go out on the road again.

Thus in 1774, at the age of forty-eight, Hutton set off for an extensive geological tour of England and Wales.

12

'Lord pity the arse that's clagged to a head that will hunt stones'

The first leg of Hutton's summer jaunt was a trip to Birmingham in the company of James Watt.

The year 1774 was a pivotal time for Watt. Roebuck, his original financier, had gone bankrupt, but Watt had found a new partner in Matthew Boulton, a manufacturer from Birmingham who had in fact been one of his creditors. In September 1773 his wife had died through complications in pregnancy. It was time for a new start, and in Boulton he found a strong supporter, 'an iron chieftain' as Boswell would call him. Watt agreed to emigrate to Birmingham, taking his Kinneil prototype engine with him. He tried to persuade Hutton and Black to travel with him, but in the event Hutton was the only one of those bachelor philosophers to make the trip.

On 17 May 1774, then, Hutton and Watt set off south.

Hutton and Watt arrived in Birmingham, after 'a pleasant journey in which nothing remarkable happened', as Watt wrote. But on the way they probably visited salt mines in Cheshire, where Hutton made a startling and satisfying observation. In one mine, the rock salt was in a layer 'thirty or forty feet deep' sandwiched in strata of red marl (a limestone clay). The salt was solid and in places pure and transparent, but a lot of it was stained with the marl.

If the salt had been laid down from water, then Hutton would have expected to see strata in it. And when he looked at the base of the rock, at first that was what he thought he saw. But further up, as Hutton reported in 1785, 'the most beautiful and regular figure was to be observed ... It was all composed of concentric circles; and these appeared to be the section of a mass, composed altogether of concentric spheres.' The salt mass was like one gigantic onion – or a pearl, or a gobstopper – with spherical layers one over the other, clearly marked out by the marl. This couldn't be the result of deposition from water: the only possible conclusion was that the whole salt mass had once been molten, and had solidified in place. And the only agent that could do that, of course, was heat. It was a remarkable, if serendipitous, bit of evidence for the role played by heat in the formation of rocks.

Hutton stayed in Birmingham for some time. He met Matthew Boulton and members of the industrialist's circle, and he used Birmingham as a base for some geological explorations of Derbyshire. Watt, meanwhile, was welcomed into the group that would become, in 1775, the Lunar Society. This group of distinguished Birmingham-based scientists and artists met monthly, on the occasion of the full moon, to advance the sciences and the arts – not quite an Edinburgh club perhaps, but congenial company. Joseph Priestley and Josiah Wedgwood were Lunaticks, and on this visit Hutton met Erasmus Darwin, James Keir, and John Whitehurst, the geologist.

Hutton no doubt argued happily with Whitehurst over his view that the present disordered state of the Earth must be due to a catastrophic pulse of heat from the interior, which had elevated the Alps and the other mountains, and left the strata in ruins. Hutton would have disagreed with much of this theory, such as its lack of a system of renewal, but at least Whitehurst was thinking about inner heat. and he did speculate on an igneous origin for basalt.

Captain James Keir, meanwhile, had also studied medicine at Edinburgh. After spending eleven years in the army he became an industrialist, with shares in glass and chemical works. Keir also dabbled in experimental geology. In 1776 he would report on experiments on the crystallisation of glass to the Royal Society in London, backing up Hutton's notions of subterranean heat: 'Does not this discovery, of a property in glass to crystallise, reflect a high degree of probability on the opinion that the great native crystals of basalts, such as those which form the Giant's Causeway ... have been produced by the crystallisation of a vitreous lava ...?' This work would influence the researches of Hutton's friend Sir James Hall twenty years later.

Erasmus Darwin, a few years younger than Hutton, would become grandfather to Charles. He was a very successful doctor – so much so that George III had offered him a position as his personal physician in London, a post that Darwin refused, not wanting to move away from his practice in Lichfield. He was a polymath like Hutton, with learned opinions on a wide variety of subjects. Something of a radical, he often wrote up his scientific opinions in verse. He would develop his own theory of evolution, though unlike his grandson, Erasmus believed that species shaped themselves to the world in a purposeful way, rather than through the random workings of natural selection.

It may be that Darwin influenced Hutton's thinking about nature, for in his final publication, Hutton would produce speculations that look in retrospect very like Charles Darwin's theories of natural selection: 'In concerning an infinite variety among the individuals of that species, we must be assured that, on the one hand, those which depart most from the best adapted constitution will be most liable to perish while on the other hand, those organised bodies which most approach to the best constitution for the present circumstances will be best adapted to continue in preserving themselves and multiplying the individuals of their race.' Unlike Charles Darwin, however, Hutton did not follow the logic of this tantalising remark to its conclusion. Conversely, Hutton left Darwin with a taste for geologising.

On this first visit, Hutton joined with Darwin in a playful experiment in which an air gun was fired onto the bulb of a thermometer, to study the cooling effects of expansion.

This congenial stay couldn't last forever, though: in July Hutton set off for some solo geologising in the wilds of Wales.

It is hard for us now to grasp the hardship that a journey like this would have meant for the forty-eight-year-old Hutton. It was a time when long-distance journeys were undertaken on foot, horseback, or – if you were lucky – stagecoach. In 1750, a weekly coach service had begun between Edinburgh and London, taking five days, cut to a mere three by the end of the century. Benjamin Franklin, during his visits, was surprised how good the English roads were. He managed to complete one hundred kilometres even on a 'short winter day', for there were stations with fresh horses every fifteen or twenty kilometres. The Welsh roads, on the other hand, were a good deal worse.

Leaving Birmingham, Hutton passed through Dudley and Stourbridge. On his first night alone he stayed in Bridgnorth, on the Welsh borders. He wrote to Clerk-Maxwell that he was looking to Wales 'with the eye of faith', 'with my arse to the east & face to the Irish channel being willing to see through Wales or at least to look at it; it will cost me some leather no doubt.' Hutton's stay with the Lunar Society folk had been stimulating, but Bridgnorth wasn't so exciting. 'I have eat green goose, but notwithstanding the weather would seem to favour spontaneous generation a slice of cucumber is all I have got in the vocable of C, and that you know is no provocative I have just muddled with brandy & water & so to bed.'

This bawdy stuff, two centuries out of date, takes a bit of translating. Hutton is really telling Clerk-Maxwell about his amatory adventures. The 'C' probably means 'cunt'. 'Green goose' was slang for a bird of less than four months old, probably meaning here a young girl. But all that seemed to be on offer in Bridgnorth was 'a slice of cucumber', meaning a married woman, which he didn't find quite so stimulating. (So much for Playfair's portrait of a sober ascetic!)

From Bridgnorth, Hutton set off into central and south Wales. In Wales he mostly had to ride on horseback, and his backside took a good deal of punishment: during this forty-day tour his riding breeches would wear out four times. He felt very sorry for himself: 'Lord pity the arse that's clagged [attached] to a head that will hunt stones.' Hutton's letters make eighteenth-century Wales sound like the Wild West: 'Upon the Holyhead road is nothing to be met with but either vultures or bad doing ... If you travel in a machine they pick the inside of your purse if on horseback you peel the outside of your arse.'

One object of his trip to Wales was to figure out the origin of a certain hard gravel of granulated quartz, much in evidence around Birmingham. He had some trouble finding the rock he wanted. But he was also looking for samples for Watt's interests in the Glasgow pottery company, and for Keir's glass works. He would write to Watt, 'I have found the crystalline, but where I least expected – I could not help singing all that day long – what do you think was the song – "she sought him east, she sought him west, she sought him broad and narrow ..." ' This was a verse of an old Scots ballad. Most importantly, in the 'primitive' Welsh mountains he was looking for traces of fossils, which would be a cornerstone of his hoped-for debunking of the Flood theorists.

By the end of July, Hutton had left Wales behind, none too impressed. He travelled through Wiltshire, and eventually reached Bath. That wasn't much fun either. The houses in the newly built and very fashionable parts of the town struck him as lacking in character: Bath was like 'a warehouse where towns may be furnished with fine places ready made'. Still, Hutton made the best of it. He stayed at the Pelican Inn in Walcot Street, just off the London Road. The Pelican was an inn of quality where Samuel Johnson would stay in 1776, but Hutton was 'as solitary as if upon the top of Mount Ararat, not a soul to speak to'.

Everywhere he went, Hutton collected geological samples, often sending them home to the safe keeping of a cousin of his partner Davie. 'I have this day packed a hogshead of bibles all wrote by God's own finger,' he wrote to Clerk-Maxwell from Bath; 'they are to go by the way of Greenock if they can find the way.' Sometimes the wait was trying; he wrote to Watt in 1774, 'Set all the Bells and hammers of Birmingham a ringing for my treasures are arrived in the Firth [of Forth] though as yet not come to hand.'

Hutton was a gregarious man who missed the companion-ship of his friends. Perhaps these days of isolation brought back memories of his lonely years on the farm. He finished one solitary evening, as he wrote to Clerk-Maxwell, by drinking the last of his 'sixpenny'th of toddy to omnibus friendibus concubinibus ubicumque' – his pidgin Latin meaning 'Here's to all concubines wherever they are.'

For Watt, meanwhile, Boulton's reliable financial input had suddenly made fast progress possible. Watt was trying to develop a new rotary steam engine as an alternative to Newcomen's venerable vertical design: the idea for a 'steam wheel' had been covered in his 1769 patent, but he hadn't had the chance to develop the concept until now. It was a key moment in the history of technology, and Hutton was keenly aware of what was happening. He wrote to Watt, 'Is your egg

hatched yet, or are you still sitting brooding like a bubly jock?'
(that is, like a turkey). And he wrote to Clerk-Maxwell of
Watt's successful experiments with his new 'curious wheel':
'This will raise his fame yonder it being so new a thing for that
is what catches the multitude ... Tell Dr Black of Watt's
success.'

Hutton had considered going on from Bath into Cornwall, but
'my money will not hold out'. (He never would explore
Cornwall.) He decided instead to go back to the Midlands: 'I
begin to be tired speaking to nothing but stones and long for a
fresh bit of mortality to make sauce to them.'

Shropshire was already a hub of the Industrial Revolution. In
the year of Hutton's visit the first iron bridge was being cast at
Coalbrookdale, to be erected at Ironbridge in 1779. Here
Hutton climbed the Wrekin, a peak some four hundred metres
high in the Shropshire Hills. The Wrekin is a mass of rhyolites
and quartzites, and its appeal for Hutton was probably related
to his interest in raw materials for porcelain manufacture.

His companion on this climb was Charles Francis Greville,
then aged twenty-five, a keen mineralogist. Greville would go
on to hold many political offices, and would also become the
protector of Emma Hart. Emma would marry Greville's uncle,
Sir William Hamilton, after Greville sent her to Naples in
1786 to be Hamilton's mistress. This was part of a deal made in
return for Hamilton paying off Greville's debts. This wouldn't
do Hamilton much good, however, as Emma became the mistress
of Horatio Nelson. After Nelson's death she squandered her
inheritances from both husband and lover, was imprisoned for
debt, and would die in poverty and exile.

The Wrekin adventure was hazardous, as the two men
climbed the summit by night and 'groped our way home two
or three miles by the light of the stars'.

After Shropshire, Hutton went back to Wales, this time to
Anglesey and the north. Anglesey is an island of low and fertile
plains, contrasting with the mountainous Welsh mainland.
Hutton would compare the straits between Anglesey and
Wales with those between Sicily and Italy as an example of
erosion; the sea had evidently broken through a neck of land
that had once joined island to mainland. On this visit Hutton

was disappointed in his quest for an unspecified 'white stuff', presumably another mineral of interest for an industrial enterprise concerning himself and Watt.

Whilst in Anglesey, Hutton visited the Parys copper mine, near Cerrig y Bleiddie in the north-east of the island. He was impressed with Parys – 'the mine is an immense lump of pyrites iron mixed with copper' – and with its working. He compared it to the immense open-cast copper mine at Fahlun in Sweden, and described the mine's processes in detail to Watt: 'The poorer ore is kindled in great heaps like lime kilns and burns of itself with the smell and quality of hell fire for three months sometimes.' The mining at Parys had begun in earnest in the 1760s, as the ships of the British navy began to have the bottoms of their hulls sheathed in copper. Parys would subsequently outstrip Cornish mining and dominate the copper trade for twenty years, and the competition it introduced would open up markers for Watt's engines, but by the 1790s production was falling, and in 1794 Hutton would be sad to hear the mine was failing.

Hutton's journey home was erratic. From Anglesey he went back to Manchester, and detoured south to see Watt. In Manchester he visited Sir Ashton Lever at Alkrington Hall. Lever had amassed a famous collection of tribal artefacts, shells, birds and fossils. Sadly the show was over by the time Hutton arrived: Lever had packed it all off to London, where it would become a notable diversion, and he would later sell it off to resolve financial problems.

After Manchester, Hutton had a good time with friends in Warwickshire, and further north: 'I made a lucky escape from Warwickshire,' he told Watt, 'but in avoiding Scylla I fell into Charibdes I was as tired of eating and drinking as a bachelor is of fasting and a married man of kissing at home.'

In the same letter, Hutton joked about growing homunculi 'as eggs in turkey & fir trees in a nursery', and developed elaborate metaphors concerning machinery and sex. This kind of stuff was common in lewd writings of the time. The Beggar's Benison members were fond of the idea of 'Merryland', an imaginary island that derived from English pornography, a marvellous place whose hillocks, shrubs and grottoes provided plenty of resource for sexual allusion. Hutton's letters show he was thoroughly immersed in his era's culture of sensuality: again, so much for Playfair's upright citizen.

Hutton travelled next to Buxton, home of Erasmus Darwin, where he was mortified to find that if not for 'the beastliness of gluttony and the manliness of drinking' he might have met Omai, who had been there just a week before. Omai was a young man who had been brought back from Tahiti after Cook's second voyage: he toured the country in the care of Sir Joseph Banks, causing a sensation in polite society.

Desite these diversions, Hutton was glad to get back at last to Scotland: 'I have undergone the most amazing hardships all last summer & harvest but at last thank God surmounting every difficulty the devil has accumulated in the way I arrived at the blessed place of my nativity.' But no sooner had he arrived than he was summoned by 'an incendiary letter' from a 'Madam Young', and off he set again to visit her at her home in the north of Perth.

Hutton had been intrigued by experiments being performed in Perthshire by Neil Maskelyne, the Astronomer Royal. Maskelyne spent the second half of the year camped next to a mountain called Schiehallion, the most symmetrical in Britain. By measuring the deviation of a plumb line on the north and south faces he hoped to prove that gravity acted here on Earth as well as between celestial bodies – the bob was attracted by the gravity of the mountain's mass. It had become a fashionable adventure in polite society to visit the great man's lodge. But though Hutton was in the area, and was interested in the project – 'the gravitation of a mountain has this summer been ascertained by Masculine [sic] and the philosophers here in the North they found it 6 minutes on the one side & on the other is 12 and they could perceive half a minute so that there is no room for error' – he devoted his attentions to Madam Young.

Madam Young's relationship to Hutton is not clear. She may have been the wife of George Young, once Hutton's medical tutor, or perhaps of his son Thomas. But polite acquaintances don't send you 'incendiary letters', and keep you from popping over to see the Astronomer Royal. Something of Hutton's complicated but secretive personal life is surely showing itself here.

Hutton finally returned to Edinburgh by late October 1774. Back home, he missed absent friends, notably Watt, to whom he wrote: 'I am returned to the empty place where my friends and yours should be ....' It had been quite a tour. In the course of a complicated six-month journey he had travelled to Birmingham, through south, central and north Wales and Anglesey, through Wiltshire to Bath, and to Warwickshire and Derbyshire. This tour is Hutton's best-documented field trip, thanks to the preservation of some of his letters in the files of James Watt (Watt had invented a copying machine for preserving his own correspondence). By comparison, Hutton left no 'memorandum' at all of his 1764 jaunt in the Highlands with George Clerk-Maxwell, for instance, as Playfair noted.

For all the exertion, he was pleased with the outcome of his tour. 'I think I know pretty well now what England is made of [save for] only a bit of Cornwall,' he told Clerk-Maxwell. He had amassed much useful and specific data on basalts he found in different parts of the country: 'The whinstone of Scotland is also the same with the toadstone of Derbyshire, which is of the amygdaloides species; it is also the same with the ragstone of the south of Staffordshire, which is a simple whinstone, or perfect trap.'

And during his trip to Wales, he had succeeded in finding 'the mark of shells ... in what may be primitive mountains'.

Hutton's fieldwork was crucial to the development of his ideas: his 'theorising would never even have begun without what he perceived in the rocks. Some modern commentators have tended to dismiss Hutton as an 'armchair geologist', spinning elaborate theories from the smoky comfort of the Edinburgh clubs. But between 1750 and 1788 he would journey through nearly every part of Britain, except Cornwall, the Hebrides and north-west Scotland, and as early as 1764 his forays into the field had become directed and specific, as he sought evidence to support his arguments. It is true that Hutton's theory was derived *a priori* from design arguments, and even in his own time he would be criticised for presenting insufficient evidence for his theories. However, he clearly saw his fieldwork as a key component of his study. To Hutton, exploration in the field was well worth a few pairs of breeches, and some solitary nights.

But if Hutton's theories were gradually being uplifted into coherence and plausibility, his life and those of his Edinburgh friends were about to be disrupted by great historical forces: there was trouble in the colonies.

13

'Britain derives nothing but loss from the dominion'

Benjamin Franklin was well known in Edinburgh. He had received a doctorate from St Andrews in February 1759, and in 1771 he paid a three-week visit to Edinburgh where he stayed with David Hume. It is highly likely that he met Hutton on this visit. But none of Hutton's circle could have guessed at the immense historical responsibility soon to be thrust upon their erudite friend.

Over the last few decades there had been a series of wars in Europe and overseas as the great powers sought to gain advantage over each other and to express their empire-building ambitions. British and French troops had been clashing for years in the American Midwest.

The Seven Years War of 1756 to 1763 was yet another messy pan-European conflict in which several powers tried to restrict the ambitions of Prussia's Frederick the Great. The difference this time was that the Seven Years War ended with France and Spain ceding most of their American territories.

The French immediately turned the loss to their advantage. Just as they had exploited the threat to the British of the Bonnie Prince, they now foresaw that as the American colonists no longer needed British protection against a French presence to the north, they would sooner or later start agitating for independence. As early as 1764, the French began to send agents to the colonies to help foment unrest. That unrest intensified faster than anybody had anticipated, thanks to the cack-handed governance of King George III and his ministers. The Boston Tea Party was one outcome of their foolish and arbitrary taxation policies.

In Hutton's circle there was dismay and foreboding. In 1775, Adam Smith made a devastating and prescient critique of Britain's handling of the issue: 'There are no colonies of which the progress has been more rapid than that of the English in

North America,' but thanks to its monopolistic policies 'Britain derives nothing but loss from the dominion.'

The unrest turned violent in April 1775, when British troops marched into Lexington, Massachusetts, to find a hundred local militia men drawn up to oppose them. With blood spilled, the uprising spread quickly.

Benjamin Franklin had been central to the diplomatic and legal manoeuvrings of the early 1770s. After being gratuitously humiliated before the Privy Council in London, however, Franklin's mood hardened against compromise. He returned to America, where he helped draft the Declaration of Independence of 1776. In 1778, he proceeded to France. Britain's long-term enemy had soon signed alliance treaties with the United States: it was the first time the new nation had been recognised as an independent state. At this key moment, Franklin wore the same velvet cloak he had worn before the hostile Privy Council in London; his revenge was cold.

The British government tried to respond: Hutton's close friend Adam Ferguson was sent to America as secretary to a Peace Commission to negotiate 'Conciliatory Proposals'. Ferguson was another relative of Joseph Black (his mother was Black's great-aunt; later Ferguson married Black's niece). Ferguson would make significant contributions to the discipline of sociology, but his early life was rather more colourful. In 1745 he had joined the Black Watch as a chaplain. This experience profoundly altered Ferguson's perspective on the Highlanders: while others saw them as uncouth barbarians, Ferguson recognised their sense of honour, loyalty, courage and generosity, and compared them to Homeric warriors. Wilhelm Friedrich Hegel would later incorporate many of Ferguson's ideas into his own analysis of history, which would subsequently be developed by Karl Marx.

The Peace Commission, however, turned out to be a disastrous fiasco, overtaken by the fast pace of events in revolutionary America. Besides, Ferguson himself wasn't an easy person to make peace with: he even fell out with his old friend Adam Smith, reconciling only when Smith was on his death bed.

American independence was finally sealed when Cornwallis surrendered at Yorktown in October 1781. The news of this remote disaster took thirty-seven days to reach London.

Scotland had sent many settlers to America, and Hutton's circle, like the rest of the nation, was divided by the news of the loss of America. Many of Hutton's friends in the merchant class, concerned about their material self-interest, backed the British government, and even offered to help raise volunteer regiments to help put down the rebellion. But David Hume's sympathies had been entirely with the colonists: he thought they should be allowed to govern themselves as they saw fit, and like Adam Smith he believed that imperial possessions would ruin Britain, both financially and morally, unless the government turned in the direction of free trade.

Hume's writings – primarily through James Madison – had influenced the drafting of the American Constitution. But Hume himself did not live to see America's independence; he died in 1776. Adam Smith wrote of him, 'Upon the whole, I have always considered him, both in his lifetime and since his death, as approaching as nearly to the idea of a perfectly wise and virtuous man, as perhaps the nature of human frailty will permit.' James Boswell, visiting Edinburgh at the time of Hume's death, inspected his open grave, and watched the funeral from behind a wall. He decided to go to the library to study some of Hume's writings as a mark of respect – but on the way he was distracted by an encounter with a young lady, and the great man's works were left, for that day, undisturbed.

Despite the turmoil, Hutton's life in Edinburgh had become settled and happy. Now in his fifties, he would dine on Sundays at Adam Smith's, and fortnightly at the home of Lord Monboddo. Monboddo was a judge some years older than Hutton who would become an early supporter and friend of Robert Burns. William Smellie was another regular at Monboddo's dinners. The son of a stonemason, by 1771 Smellie had become the editor of the first edition of the *Encyclopaedia Britannica*, a project he approached with 'a pair of scissors' and considerable acumen.

By 1778 Hutton had become well enough established in Edinburgh circles to set up his own gentlemen's club, the Oyster Club. He, Joseph Black and Adam Smith were its founder members, and in the years that followed the club became very popular. At its weekly meetings, over tankards of porter and feasts of oysters – a fashionable delicacy – 'the conversation was always free, often scientific, but never didactic or disputatious; and as this club was much the resort of the strangers who visited Edinburgh, from any object connected with art or with science, it derived from thence an extraordinary degree of variety and interests.' Members of a new generation were among the luminaries of the Oyster Club – including Hutton's companions at Siccar Point a decade later.

Sir James Hall was the son of Sir John Hall of Dunglass, Hutton's old friend from his farming days. Hall acceded to his father's baronetcy on the latter's death in 1776, becoming Sir James as a boy of fifteen. After studying at the university under Black and others, Hall became an able chemist, and would be fascinated by the notion of testing geological ideas in the laboratory. In the 1780s, following the custom of the time, the young Hall undertook a Grand Tour of Europe, during which he investigated the great volcanoes of Italy, including Vesuvius, then actually erupting. In Edinburgh's National Portrait Gallery you can see a portrait of Hall painted in Rome at the end of this tour: the twenty-five-year-old has a broad, determined face and lively eyes under a mane of prematurely receding blond hair. It is quite a contrast to the portrait of Hall in old age which hangs in the lobby of the Royal Society of Edinburgh, in which the great scientist broods ferociously over a book. Thirty-five years younger than Hutton, Hall would neverthe-less become a close friend – surely a mark of Hutton's warmth and lack of envy.

John Playfair was born in 1748: the son of a minister from Dundee, he was educated at home by his father, and then sent to Edinburgh to qualify himself for the church. Mathematics proved more conducive to Playfair's mind than religion, but when his father died in 1772, Playfair succeeded to his father's parish in Dundee. Ten years later, however, he gave up his church livings. For a time he made a living as a private tutor, and in 1785 he would be appointed professor of mathematics at the University of Edinburgh. Playfair would gain a reputation in mathematics, the history of science and geology – but his chief claim to fame is as Hutton's Boswell, translat-ing (and sometimes refashioning) his friend's thoughts for later generations.

The Rev. William Robertson, an intimate of Hume's, was another pearl of the Oyster Club. Once one of Maclaurin's volunteers in the 'Forty-Five, Robertson became 'Historiographer Royal', led the Moderate Party in the General Assembly of the Church of Scotland, and was principal of the University of Edinburgh for thirty years. He was a talented writer: his histories of Charles V, America, India and other subjects won plaudits from such luminaries as Voltaire and Catherine the Great. Hutton would later turn to Robertson for advice on the drafting of his first formal presentation of his geological theory.

Adding to this rich mix was a Russian princess. Catherina Romanova Vorontsov Dashkov was a friend of Diderot and Voltaire. She spent some time in Edinburgh, living at Holyroodhouse, and described the Edinburgh luminaries as 'esteemed for their intelligence, intellectual distinction and moral qualities'. There were many links between Russian and British scientific circles: in 1783 Joseph Black would be elected an Honorary Member of the St Petersburg Imperial Academy of Sciences, and in 1818, the Russians would try to recruit the geological map-maker William Smith to supervise the immense coal industry in the Ukraine.

Meanwhile, Watt's career had blossomed. In 1776 the first two Watt engines were installed to pump water for a Staffordshire colliery and to blow air into the furnaces of ironmaster John Wilkinson. Watt continued to improve his designs. He introduced the 'double-acting' engine in which steam was applied alternately to both sides of the piston, thereby doubling the power obtainable from the same size of cylinder. And in 1782 he took out a patent for a new engine in which a rocking beam drove a shaft and flywheel: Watt now had an engine capable of driving machinery. In the mines, everything would change: lifts could now lower workers to the coalfaces and haul their coal back to the surface again.

Watt's breakthroughs gave him a monopoly over British steam-engine design for a quarter of a century. At last he was doing well financially, and in 1785 he would be elected a member of the Royal Society in London. Watt always retained his contacts with Edinburgh, however. He wrote to Black: 'Everybody here who knows [you] would be happy to see you and I hope you will bring Dr Hutton with you, perhaps we may find some stone for him and our iron works would please you both.' But Hutton was too settled in Edinburgh; he did not visit Birmingham again. In 1778 he did receive Erasmus Darwin as a visitor, but it was not a happy occasion. Darwin's son Charles, another Edinburgh medical student, had died of an infection from a cut finger. Darwin stayed with Hutton until after the funeral.

Hutton had also kept busy with many commercial projects. His sal ammoniac factory was still in operation, and diversifying: in 1783 it would begin to process crude sal ammoniac bought in from a tar works at Culross. Hutton also experimented with varnishes. There was a great demand for varnish to protect ironwork and for domestic articles and ornaments: Watt's financial partner Matthew Boulton turned out a variety of varnished fancy goods – which Hutton called 'beads, jinkumbobs & little Jesuses'. The beads were much valued by explorers like Cook as trade goods. In Birmingham and Wolverhampton the manufacturers of an imitation oriental lacquer were experimenting with coal tar as a base, but this caused noxious fumes. Hutton tried using coal rather than coal tar in this process, and corresponded with Watt over the issue. In this enterprise at least he seems to have been unsuccessful.

Whatever other interests he pursued, Hutton always continued to geologise.

He had no formal connection with the university, but would freely use his rock collection to assist enthusiastic students. One wrote of him, 'He is an excellent mineralogist, and is very communicative, very clear, and of a candid though quick temper. He has a noble collection of [rock samples], which he likes to show.' Another of these students had Christmas dinner with Black and Hutton: '[Black] himself is so lazy he is obliged to get Dr Hutton to be master of all the ceremonial part. The Doctor likes to sleep after dinner.'

Hutton also showed his collection to visiting scholars. But this led him into a somewhat embarrassing confrontation with the prominent French geologist Barthelemy Faujas de Saint-Fond, who toured Scotland in 1784.

Faujas had been a lawyer and a commissioner of mines before

becoming geology professor at Paris's natural history museum. He visited the Western Isles, where he studied columnar basalts at Fingal's Cave, comparing them with similar formations in France. In Edinburgh Adam Smith invited the Frenchman, known for his love of music, to a bagpipe competition. Faujas described how 'the competitors ... formed themselves into a line two deep, and marched in that order to the castle of Edinburgh, which is built on a volcanic rock' (as indeed it is).

Hutton proudly showed off his rock collection to the visitor. But Faujas noted that the samples lacked the rock matrices in which they had been found, so reducing their value. 'I therefore had much more pleasure in conversing with this *modest philosopher* than in examining his collection' (my emphasis). The snooty Gaul was more impressed, in fact, by Black, who he called a 'learned chymist'.

Modest philosopher or not, by now Hutton was doing some very deep thinking about his geology. Throughout his Edinburgh years the core of his theory of the Earth had never wavered in Hutton's mind. He had sought to flesh it out with new data and evidence, which he had found in the field, and with theoretical concepts that he found in the prolific thinking of his friends, notably Black, Watt and Hume.

Hume's work inspired Immanuel Kant's critical philosophy, and was one of the influences that led Auguste Comte to positivism. In Britain, he would influence Bentham, Mill and others. Hume's empiricism, proposing that all knowledge must be based on experience, and his scepticism, holding that we must always be aware of the discrepancy between what we claim to be capable of and what we are, would have a great influence on later generations. His overriding lesson for thinkers like Hutton was that scientists, even those occupied by an apparently 'concrete' discipline like geology, need to be careful not just about what they claim to know but also about *how* they claim to know it. Human reason is a fragile thing and prone to be overthrown by suggestibility.

It was a lesson Hutton learned very well. Hutton was wary of the wild and undisciplined speculation with which those who had become known as 'system-builders' had tarnished geological thinking. As Playfair wrote, such geologists were 'men often but ill-informed of the phenomena which they proposed to explain, and who proceeded also on the supposition that they could give an account of the *origin* of things … Men who guided their inquiries by a principle so inconsistent with the limits of the human faculties, could never bring their speculations to a satisfactory conclusion.' This had rendered Burnet's schema, for example, no more than a 'poetic fiction'.

If Hutton was to avoid being accused of the same groundless confabulating, he came to see, it wasn't enough just to set out his geological knowledge: he would have to demonstrate how he had obtained his knowledge about places and times unseen – and Hume's empirical, observation-driven philosophy, fused with the Newton–Maclaurin tradition, would be crucial to this project.

Hutton's geological ideas, however, remained far from the contemporary mainstream. By the 1780s, the ideas of Abraham Werner dominated European geology. Werner was a practical man and a good teacher: a professor at a mining academy in Saxony, he attracted students from all over Europe. He published little on his theory, but his lectures, which he began to give in Freiburg in 1775, were immensely influential, and soon his students were taking his message across the Continent.

And Werner, like many of his predecessors, believed in deposition from a primordial ocean.

Werner argued that the Earth had originally consisted of a primeval nucleus completely enclosed by some universal liquid – perhaps water. Earth's rocks precipitated out of this fluid; as time passed the composition of the fluid somehow changed, and hence the nature of the rocks differed, one layer on top of another – which explained the orderly layering of the strata now being mapped out in the field.

The earliest 'Primitive' rocks were granite, and also gneiss, schist and quartz, and, since they were formed in a period before life, they were devoid of fossils. Next the fluid levels lowered and the continents emerged. The rocks formed in this 'Transition' era were a mixture of sediments and precipitates and therefore contained some fossils. In the final, modern era, much of the fluid had drained away, leaving the modern oceans. Rocks formed in this period, the *'Floetz'* – like mudstone and sandstone – were created by erosion and were full of fossils.

The last layer of all was alluvial, the loose deposits of gravel, sand and clay formed from the most recent erosion of older rocks.

Fluid was so important in this argument that the theory was known as 'Neptunist', after the god of the sea. Even basalts were considered to have crystallised from the primordial fluid. Only those few rocks, lavas and volcanic ashes that were known by observation to have erupted from volcanoes, were thought to have originated from subterranean fires – which were believed to be burning banks of underground coal.

All this was familiar to Hutton. Many of these ideas had been prefigured by François Rouelle, who may have taught Hutton in Paris, and who had first proposed that granite was a primitive rock precipitated out of a universal ocean. Werner's theory had all the old problems – such as, where had all the primeval fluid gone? But it made a firm prediction, that the same kinds of rocks should have been laid down in the same sequence all over the world, and this had encouraged geologists to make careful studies of the rocks. Out of this would come the first attempts at 'stratigraphy' – that is, mapping the Earth's strata as they were actually to be found in the ground, and trying to figure out their history.

Meanwhile, there had been increasing disquiet about the Earth's true age. As the real history of the Earth was uncovered, fragment by fragment, by the patient work of field geologists, it became increasingly difficult to imagine that the many mighty events documented in the rocks could all have been crammed into the brief Ussher time frame. The geologist Charles Lyell, writing in 1830, would say this was like trying to squeeze the events of a thousand years of human history into a mere century: 'Armies and fleets would appear to be assembled only to be destroyed, and cities built merely to fall into ruins.'

Werner himself did not buy the Ussher chronology. He was actually a Deist whose thinking mirrored Hutton's to some extent: the Earth's rocks were the product of rational, goal-oriented processes that showed the working of God's hand, but the literal descriptions of Genesis could be ignored. He guessed it was perhaps a million years ago that his primeval ocean covered the Earth. But Werner kept such thinking private; the whiff of atheism still hung around any attempts to dispute Ussher.

By 1785 Werner's ideas had dominated for a decade – and he was twenty-three years younger than Hutton. There was still no viable theoretical alternative to their ocean-deposition-and-decay model of Earth's history.

None apart from Hutton's, that is.

Having toyed with his theory of the Earth for perhaps twenty-five years, Hutton was now nearing sixty – but he had still to present his theory formally, or even properly to write it down. According to Playfair he had communicated it only to his friends Black and Clerk of Eldin, and 'though fortified in his opinion by their agreement with him (and it was the agreement of men eminently qualified to judge), yet he was in no haste to publish his theory; for he was one of those who are much more delighted with the contemplation of truth, than with the praise of having discovered it.' He was not a professional academic, and so there was no *need* to publish. And perhaps Hutton, knowing how far he had strayed from the prevailing paradigms, feared for the future of his precious theory once it was exposed to the full glare of scrutiny, beyond the cosy circles of his Edinburgh friends.

Now, though, he came under pressure to publish for the first time.

In the 1780s the Scottish Enlightenment was at its peak. World-renowned figures thronged Edinburgh's clubs and societies, and the classes at its universities in medicine and science were among the best in the world; by the end of the century Edinburgh had more students than Oxford and Cambridge put together. Now was the time, it was decided, to set up an enduring intellectual monument to this golden age.

All around Europe, new institutions devoted to the sciences and the arts were being established. These could be traced back to the Italian Academies of the sixteenth and seventeenth centuries. The Royal Society of London had been founded in 1660, and on 1 March 1665 it had begun the publication of its *Philosophical Transactions*, a series which continues to the present day. Shortly, the Royal Irish Academy and the French Academy of Sciences were to be established. But, a century after London, what of Edinburgh?

It would be nice to report that the founding of the new Royal

Society of Edinburgh was the outcome of civilised and sedate discussions. In fact it was born of jealousy, turf warfare and bitterness; even founders of Royal Societies are human.

It seems entirely appropriate that a nucleus for Edinburgh's new institution should have been a gentleman's club: the Rankenian, formed in 1716 for 'literary social meetings' held in Ranken's Inn. The Rankenian lasted sixty years, and counted Colin Maclaurin among its alumni. Maclaurin, of course, had also been responsible for establishing the Philosophical Society in 1737. But the Philosophical Society was a small, not well-organised voluntary body, and with the passing of time it had become lifeless and uncertain. There was clearly a gap to be filled.

In 1782, the recently formed Society of Antiquaries of Scotland stepped into this vacuum. The antiquarians proposed to seek a Royal Charter, to cultivate knowledge not just of antiquities but of other areas, including natural history; they even proposed setting up a natural history museum. This provoked an instant reaction from the University of Edinburgh, who feared that the proposed lectures would rival its own, and from the Faculty of Advocates, who worried that the new museum would compete with its own library, which housed antiquarian objects as well as books and manuscripts.

John Walker, professor of natural history at the University, was particularly concerned about the antiquarians stepping on his turf. In that same year of 1782, he drew up a 'proposal for establishing at Edinburgh, a Society for the Advancement of Learning and Useful Knowledge'. The University, the Faculty of Advocates, and the Philosophical and Antiquarian Societies would unite to form a Royal Society of Edinburgh under a Charter from the King. Walker's tactic, aimed at protecting his own professional interests, was to bring the antiquarians to heel. The antiquarians would have none of it, and after three months of acrimonious meetings they left the proposed coalition. But by the end of the year a petition had been sent to London, by 29 March 1783 the King's signature had been secured, and Walker had achieved his aim.

The new Royal Society of Edinburgh rapidly organised itself. The first assembly included Robertson, Walker, Adam Ferguson and Adam Smith. The fellows of the old Philosophical Society were subsumed as fellows of the Royal Society, so that Hutton became a Founding Fellow. The fellows were organised into a Literary Class and a Physical Class, the latter including Hutton, Watt, Clerk of Eldin, Hall and others. One of the first honorary fellows was Benjamin Franklin, and the Society's first councillors included John Maclaurin, son of Colin. (John Maclaurin – who would go on to become a judge – had a less reputable side. Since 1760 he had been publishing bawdy satirical verse, beginning with *The Keekeiad*, an epic poem about a citizen of Edinburgh who set his unfortunate wife's pubic hair on fire.)

The first meetings were held in the University Library. The *Transactions of the Royal Society of Edinburgh* would soon begin to disseminate the Society's learning – although in those early days there were some puzzling omissions. A debate conducted between Hall and Hutton during 1788 and 1791, weighing the merits of the old phlogiston theory against those of Lavoisier's ideas on combustion, was never published: as the Council said in the first volume, 'Several papers have been communicated with the sole purpose of furnishing an occasional entertainment to the members, and this end being attained, have been withdrawn by the authors.'

For the Edinburgh luminaries, the founding of the Society was a statement of intellectual independence and, perhaps, of national pride: seventy-six years after the Act of Union, the Society was a much more meaningful declaration of nationhood than the Bonnie Prince's war-making.

Now, at last, prompted by his 'zeal for supporting a recent institution which he thought of importance to the progress of science in his country' – as Playfair would say – Hutton felt ready to give his precious ideas their first public presentation. But, in swimming so hard against the prevailing tide, he knew that he would have to work hard to establish his arguments.

Nervously, Hutton began to go through his notes and samples.

14

'We have now got to the end of our reasoning'

Hutton's oral presentation to the Royal Society of Edinburgh
was given in two parts, on 7 March and 4 April 1785. Hutton
was ill that March day, and the first part of his paper was read
in his absence by Joseph Black.

We must imagine Hutton's emotions as he prepared to
present his ideas before an audience of his peers, those whose
opinions mattered more to him than anybody else's in the
world. Even Playfair, who was in the audience, had not heard
a full presentation before. And we must imagine Black's precise,
clipped voice as he began, on behalf of his friend: 'When we
trace the parts of which this terrestrial system is composed,
and when we view the general connection of those several
parts, the whole presents a machine of a peculiar construction
by which it is adapted to a certain end. We perceive a fabric,
erected in wisdom, to obtain a purpose worthy of the power
that is apparent in the production of it ...'

Before he could get to the theory proper, Hutton had to tell
his audience *how* to think about geological problems: 'If, in
pursuing this object, we employ our skill in research, not in
forming vain conjectures; and if *data* are to be found, on
which Science may form just conclusions, we should not
long remain in ignorance with respect to the natural history
of this Earth, a subject on which hitherto opinion only, and
not evidence, has decided: For in no subject is there naturally
less defect of evidence, although philosophers, led by
prejudice, or misguided by false theory, have neglected to
employ that light by which they should have seen the system
of the world ...'

Hutton would not publish his theory on the nature of
knowledge until 1794, but when he did, it would become
apparent that he had worked out his own empiricism, quite

similar to that of Hume. Where Hume had talked of the difference between *impressions* and *ideas*, so Hutton split up the thought process into three stages. In *sensation* the mind receives a stimulus from some external agent through the senses; in *perception* the sensations are compared and organised; in *conception* general ideas like space, time, motion are derived from perceptions.

In the end, said Hume, all knowledge originates in experience; everything else you have to infer. And that was how Hutton had come to think about geology. Geology deals with the interior of the Earth, the depths of the sea – even the deep past and far future – all places we cannot directly observe. How can we have knowledge of something we have never examined? Just as Hume argued, we have to begin with experience, with *sensation* – or as Playfair put it, inquiries had to be consistent with 'the limits of human faculties'. We can look only at the rocks in the present, on the surface of the Earth or in the laboratory, and we can observe only the processes acting on them in the present – like erosion.

How do you infer the past from what you see in the present? 'In examining things present, we have data from which to reason with regard to what has been; and, from what has actually been, we have data for concluding with regard to that which is to happen hereafter. Therefore, upon the *supposition* that the operations of nature are equable and steady, we find, in natural appearances, means for concluding a certain portion of time to have necessarily elapsed, in the production of those events of which we see the effects' (my emphasis).

Hutton *assumed* that the same processes you see today must have been working on the rocks in the past, and will continue to do so in the future – and, building on that, you can reconstruct what must have been in the past. All you have to do is run the tape backwards from the present. Thus we can deduce that the Edinburgh hills were once the hearts of mighty volcanoes, before erosion did its work.

This is only a *conception*. You can't *prove* that physical laws never changed – perhaps the force of gravity varied a million years ago, perhaps the boiling point of water was different – but if you do assume that the same forces acting now have always acted, you have a way to unlock history. Later

elaborated by Charles Lyell under the label of 'uniformitarianism', this assumption lets us reconstruct history, and
even project the future, based on the world as we find it today
and the processes that have shaped it: 'The present is the key
to the past,' as geologist Archibald Geikie would say a century
later.

This was little understood at the time. Hutton's statement
of his epistemological thinking was too brief. Even his 1794
book on the subject, when it eventually appeared, was pretty
impenetrable, and its links to his geology were obscure: it has
taken detective work by modern scholars to make it all clear.
But this aspect of Hutton's work was crucially important. By
separating his observations from his inferences, Hutton was
trying to explain the basis on which he had derived his
hypotheses, and by opening up his methodology for examination he was setting out his thinking as a basis for a true
science of geology in the future – for that methodology itself
could be improved.

The notion of uniformitarianism is one of Hutton's key
contributions to geology. Hutton gave us more than a theory of
the Earth. He gave us a way to think about geology, a way
whose influence persists to this day.

But enough epistemology. Now, perhaps with relish, Hutton
turned to the rocks.

How were sedimentary rocks formed? You could see from
their cemented-together composition of grains and rubble
and fossils that they had been put together from the products
of erosion, and their fossil content showed they had formed
under the sea: 'If, for example, in a mass of marble, taken
from a quarry upon the top of the Alps or Andes, there shall
be found once cockle-shell, or piece of coral, it must be concluded, that this bed of stone had been originally formed at
the bottom of the sea, as much as another bed which is
evidently composed almost altogether of cockle-shells and
coral.'

It was easy to imagine great layers of rubble being formed as
the rivers washed debris into the ocean. But what could have
consolidated the strata into rock? Werner's model had all the
rocks being separated out of solution in the waters of a mighty

ocean, and that was that. But that wasn't good enough: not everything dissolves in water (or indeed, any other solvent). Rocks had been found bound together by such substances as feldspar and metals, none of which were soluble in water. Water simply wouldn't do as an explanation, unless you ascribed unobserved powers to it: 'We cannot allow more power to water than we find it has in nature; nor are we to imagine to ourselves unlimited powers in bodies, on purpose to explain those appearances, by which we should be made to know the powers of nature.'

Equally, if the rocks had been deposited out of a solution you ought to find strata being laid down in simple ways, one on top of the other. But there were some *very* peculiar rock specimens to be found in the field. Hutton had a piece of granite from Portsoy (actually brought back by John Clerk of Eldin) that contained crystalline structures of quartz and feldspar embedded *within* each other: 'The feldspar, which is contained within the quartz, contains also a small triangle of quartz, which it incloses.' So the feldspar and the quartz were nested like Russian dolls. (This complex interweaving of mineral types reminded Hutton of handwriting, and the granite type is still called 'graphic'.) How could the two mineral types have got so mixed up if they had been laid down out of a solution? Wouldn't one precipitate out first, and then the other?

But on the other hand, if the two mineral types had once been *molten*, they could easily have both solidified out of the melt in the jumbled way shown in Hutton's samples. 'The loose and discontinuous body of a stratum may be closed by means of softness and compression; the porous structure of the materials may be consolidated, in a similar manner, by the fusion of their substance; and foreign matter may be introduced into the open structure of strata, in form of steam or exhalation, as well as in the fluid state of fusion; consequently, heat is an agent competent for the consolidation of strata, which water alone is not.' Hutton turned out to be right about the heat origin of his graphic granite, although it was not until 1986 that samples were artificially produced in the laboratory, by allowing quartz and feldspar simultaneously to crystallise out of a melt.

Another piece of evidence for the action of heat was the cracks that could be observed in the strata: 'If, again, strata have been consolidated by means of heat, acting in such a manner as to soften their substance, then, in cooling, they must have formed rents or separations of their substance, by the unequal degrees of contraction which the contiguous strata may have suffered ... There is not in nature any appearance more distinct than this of the perpendicular fissures and separations in strata. These are generally known to workmen by the terms of veins or backs and cutters; and there is no consolidated stratum that wants these appearances.'

In these remarks, Hutton knew he was directly taking on Werner and his disciples. Granite, so the Wernerians claimed, was 'primitive', the oldest rock of all, the first laid down out of the universal fluid. Now Hutton was saying it was in fact one of the *youngest* of rocks, and that under the influence of great heat it had once flowed like melted chocolate. But as he developed his argument, he was in any case challenging the whole notion of primary and secondary rocks: 'We are not at present to enter into any discussion with regard to what are the primary and secondary mountains of the Earth; we are not to consider what is the first, and what the last, in those things which now are seen ...'

Hutton bombarded his audience with more examples to reinforce his case that heat, not water, was responsible for forming rocks. He referred to 'hand specimens' from his own collection: 'Here, for example, are crystallized together in one mass, *first*, *Pyrites*, containing sulphur, iron, copper; *secondly*, *Blend*, a composition of iron, sulphur, and calamine; *thirdly*, *Galena*, consisting of lead and sulphur ... All these bodies, each possessing its proper shape, are mixed in such a manner as it would be endless to describe, but which may be expressed in general by saying, that they are mutually contained in, and contain each other.' He described the rock salt deposit he had found with Watt in Cheshire in 1774, the great marl-stained onion-shell mass which showed clear signs of having congealed from a melt. He referred to nodules in the basalts of Calton Hill, one of Edinburgh's volcanic plugs. He described the chalk belts of the Isle of Wight which he had viewed during his agricultural training days, which show different degrees of consolidation.

He talked about a species of marble from Spain, in which the constituent fragments fit together as neatly as the bony plates of a skull: 'The gravel of which this marble is composed, consists of fragments of other marbles of different kinds ... Besides the general conformation of those hard bodies, so as to be perfectly adapted to each other's shape, there is, in some places, a mutual indentation of the different pieces of gravel into each other; an indentation which resembles perfectly that junction of the different bones of the cranium, called sutures, and which must have necessarily required a mixture of those bodies while in a soft or fluid state.'

And then there was basalt. Hutton described intrusions of one rock type into another. There was the monolithic slab of basalt he had first observed at Crieff in 1764. The 'dyke' (an intrusion that cut across the strata) had been left exposed and sticking up into the air when the softer rocks within which it had formed had worn away. 'It runs from [Crieff] eastward, and would seem to be the same with that which crosses the river Tay, in forming Campsy-lin above Stanley, as a lesser one of the same kind does below it ... It may be considered as having been traced for twenty or thirty miles, and westwards to Drummond castle, perhaps much farther.' Similarly, the Salisbury Crags of Arthur's Seat were a 'sill', where the basalt has run parallel to the older strata.

How could such formations occur? If sufficiently heated, rocks melt, to form magma. If magma reaches the surface it is called a lava flow – but if it is formed underground it can be forced into fissures in older, solid rocks. Hutton also quoted the practical experience of miners. Sometimes prospectors could find intrusions of the minerals they sought at *right angles* to the surrounding strata. How could these vertical intrusions have got there, unless molten material had been forced into strata already formed?

This, said Hutton, was how sedimentary rocks were formed: layers of rubble, laced with the relics of living things, were laid down on the bottom of the sea, and then the Earth's inner heat baked them into rock. Sometimes the same heat drove great veins of lava or minerals into pre-formed strata.

But how were the newly created rocks raised from the bottom of the sea, to be transformed into plains and mountains?

*

Great violence has been done to the Earth.

In some places on the planet, great successions of strata lie undisturbed. The Grand Canyon is one example. If you were to climb the walls of one of those great water-carved gullies, you would be climbing up through more than five hundred million years of geological time, neatly stacked like typing paper.

But as Hutton knew, such places are rare: Scotland certainly isn't like that. Instead, 'The strata of the globe are actually found in every possible position: Far from horizontal, they are frequently found vertical; from continuous, they are broken and separated in every possible direction; and, from a plane, they are bent and doubled.'

Then there was uplift.

Hutton had a scrap of fossil wood with an extraordinary history. The wood must obviously have come from a tree that had grown on dry land. But you could see that 'it has been eaten and perforated by those sea worms which destroy the bottom of our ships'. So after it had grown, the relic had been carried *beneath* the sea. Finally, petrified, the scrap had been raised *above* the sea once more, to finish its journey on the Isle of Sheppey, where Hutton found it. This unassuming bit of wood was evidence that the land had gone through a great cycle, of erosion, consolidation, and uplift: it had been carried through a mighty revolution in the Earth.

To Hutton this was evidence that great energies slumbered within the Earth. For another demonstration, Hutton pointed again to the existence of mineral veins running through rocks: 'Let us ... consider what power would be required to force up, from the most unfathomable depth of the ocean, to the Andes or the Alps, a column of fluid metal and of stone. This power cannot be much less than that required to elevate the highest land upon the globe.'

As the engine of uplift, Hutton settled on Earth's internal heat. The planet was a mighty heat engine, turning thermal energy into mechanical energy with the alacrity of one of James Watt's giant kettles. And the heat provided the energy required not only to bake rocks but also to uplift landscapes and mineral seams.

Hutton gave several examples of inner heat. Volcanoes were

found everywhere: 'Naturalists, in examining different countries, have discovered the most undoubted proofs of many ancient volcanoes, which had not been before suspected. Thus, volcanoes will appear to be not a matter of accident, or as only happening in an particular place, they are general to the globe, so far as there is no place upon the Earth that may not have an eruption of this kind; although it is by no means necessary for every place to have had those eruptions.'

On an Earth made for a purpose, however, volcanoes were part of the grand design. Echoing Strabo, he said: 'A volcano is not made on purpose to frighten superstitious people into fits of piety and devotion, nor to overwhelm devoted cities with destruction; a volcano should be considered as a spiracle to the subterranean furnace, in order to prevent the unnecessary elevation of land, and fatal effects of earthquakes.'

While it might be difficult to imagine the formation of sedimentary rocks, Hutton said, the fact of the subterranean heat engine ought to be obvious to all: 'To see the evidence of marble, a body that is solid, having been formed of loose materials collected at the bottom of the sea, is not always easy ... But when fire bursts forth from the bottom of the sea, and when the land is heaved up and down, so as to demolish cities in an instant, and split asunder rocks and solid mountains, there is nobody but must see in this a power, which may be sufficient to accomplish every view of nature in erecting land, as it is situated in the place most advantageous for that purpose.'

But what *was* heat? In the absence of a developed physical theory, to flesh out his geological model Hutton had been forced to develop his own ideas – ideas that would later get him into a great deal of trouble. And confusingly enough, Hutton's theory of heat, as he had eventually developed it, was neither a modern atomic theory nor a contemporary caloric one.

Hutton imagined that light, heat and electricity all came from the sun – all 'modifications of the solar substance' – and they lacked mass and weight. So Hutton's heat was *not* like caloric, which was supposed to have a weight. Rather, Hutton thought that 'The solar substance, when compounded in an inflammable body, is phlogiston; and when restored to its former liberty, to its natural motion, it is light.'

Phlogiston was a hypothetical substance which was supposed to be part of every combustible material. When you burned something, phlogiston was driven off, with the 'dephlogisticated' substance left as a residue – thus unburnt wood was actually made up of ash and phlogiston. Hutton was probably attracted to such theories because of his farming experience. He had seen plants grow in sunlight; he believed that plants used 'solar substance' in the form of light to make organic matter and phlogiston, and animals' digestion of plant material released the phlogiston, turning it back into light – a kind of mirror image of the oxygen-based physiological cycle we understand now.

However, the phlogiston theory was discredited by Antoine Lavoisier, between 1770 and 1790. He showed that burning actually involves combination with the newly discovered element, oxygen. So even by the time of Hutton's presentation, phlogiston theories were seriously old-fashioned. One of Hutton's failings, as Playfair would admit, was that he didn't always read as widely as he should have. While he devoured books on 'voyages, travels, and books relating to the natural history of the Earth', he 'bestowed but little attention on books of opinion and theory ... he was not very anxious ... to be informed of the views which other philosophers had taken of the same subject. He was but little disposed to concede anything to mere authority; and to his indifference about the opinions of former theorists, it is probable that his own speculations owed some part, both of their excellencies, and their defects.' On the issue of phlogiston, he had simply let himself get out of date.

All this was heady stuff – but how did it apply to Hutton's theory of the Earth?

Hutton imagined a *closed system* inside the Earth – something like modern hydraulic machinery. His heat, as the working fluid, circulated endlessly, melting the rocks and pushing up the land. It was just like a steam engine, with the steam circulating to push the pistons and drive the wheels.

Where had the interior heat come from? Hutton didn't know – but he believed it didn't matter, as the system was closed; the store of heat fluid would have been created with the Earth itself, and had remained unchanged ever since. He argued that the fact that he could not identify its source was not a valid objection: 'In opposition to this conclusion, it will not be allowed to allege, that we are ignorant how such a power might be exerted under the bottom of the ocean; for the present question is not, what had been the cause of heat, which has appeared to have been produced in that place; but, if this power of heat, which has certainly been exerted at the bottom of the ocean for consolidating strata, had been employed also for another purpose, that is, for raising those strata into the place of land.'

So the present-day lands had been created from the rubble of older landscapes, consolidated and uplifted by heat. But what was this great cycling world-machine *for*? Hutton's final remarks were his most visionary, as he moved to unite his arguments in a new image of the Earth on which we live.

The purpose of Earth, said Hutton, was to sustain life.

He recalled his paradox of the soil. Erosion creates soil, which is the basis of all life. On a world like Werner's, dominated only by erosion, the destruction of soil and mountains must soon bring an end to the Earth as a habitable world. But Hutton had observed a process of renovation: 'Nature does not destroy a continent from having wearied of a subject which had given pleasure, or changed her purpose, whether for a better or a worse; neither does she erect a continent of land among the clouds, to shew her power, or to amaze the vulgar man ... But with such wisdom has nature ordered things in the œconomy of this world, that the destruction of one continent is not brought about without the renovation of the Earth in the production of another.'

This was intended for the preservation of life through the creation of soil: 'A soil, adapted to the growth of plants, is necessarily prepared, and carefully preserved; and in the necessary waste of land which is inhabited, the foundation is laid for future continents, in order to support the system of this living world.'

All this needed time, of course – but how much time? 'The formation of a future Earth being in the bottom of the ocean, at depths unfathomable to man,' is unobservable. 'But, in the destruction of the present Earth, we have a process that is performed within the limits of our observation ... But how shall we measure the decrease of our land? Every revolution of the globe wears away some part of some rock upon some coast; but the quantity of that decrease, in that measured time, is not a measurable thing ...'

Hutton had claimed that the only processes that have shaped the Earth in the past are those working in the present day, such as erosion. Those effects are feeble in themselves, but Hutton assumed that 'the natural operations of the Earth, *continued in a sufficient space of time*, would be adequate to the effects which we observe' (my emphasis). That is, if you have slow-working processes, you need to allow huge amounts of time to allow them to wreak mighty changes.

A human lifespan was too short to allow precise measurements of such slow processes. Even a reference to the observation of the ancients, across thousands of years, was of no help: 'Let us then go to the Romans and the Greeks in search of a measure of our coasts, which we may compare with the present state of things. Here, again, we are disappointed; their descriptions of the shores of Greece and of Italy, and their works upon the coast, either give no measure of a decrease, or are not accurate enough for such a purpose.'

A geological cycle was *long*, enormously long – how long couldn't be determined. But if the change was difficult to measure even on intervals dating back to the ancients, time must be vastly deeper than Ussher's paltry few millennia. 'We are certain, that all the coasts of the present continents are wasted by the sea, and constantly wearing away upon the whole; but this operation is so extremely slow, that we cannot find a measure of the quantity in order to form an estimate. Therefore, the present continents of the Earth, which we consider as in a state of perfection, would, in the natural operations of the globe, require a time indefinite for their destruction ...'

And if there had been one cycle, there must have been many more. 'The world which we inhabit is composed of the materials, not of the Earth which was the immediate predecessor of the present, but of the Earth which, in ascending from the present, we consider as the third, and which had preceded the land that was above the surface of the sea, while our present land was yet beneath the water of the ocean. Here are three distinct successive periods of existence, and each of these is, in our measurement of time, a thing of indefinite duration.'

In his sometimes tangled prose, Hutton was pointing out that if the present-day mountains had been formed from the rubble of a previous cycle, then the mountains of *that* cycle must have been formed from the wreckage of an even earlier cycle – and so on.

A central theme of all Hutton's thinking about physics was that the universe was filled with two kinds of forces: attractive and repulsive. In Newton's orderly solar system, inertia was the 'repulsive' force which tends to pull a planet from its path, but this was matched by gravity, the 'attractive' force, and the balance kept the planet on its circular course. In Hutton's Earth, the attractive force was gravity, which acted to pull everything towards the centre of the planet. The repulsive force was heat, which pushed up the land. As in any well-balanced system the repulsive and attractive forces oscillated, keeping in overall equilibrium – which was why the Earth's expansive forces didn't make it balloon or explode. On the contrary: just as Newton's planets orbited in the heavens, so continents orbited through Earth's interior.

On Hutton's Earth, everything was in balance. Everything cycled; everything revolved. Hutton evoked a whole series of regenerations, of erosion being repaired by consolidation and uplift, as new lands were born from the wreckage of the old, over and over, using his heat-fuelled processes of lithification and uplift. And having no limit in time, a cycling, self-renewing Earth was surely a more perfect design than a world doomed to decay as soon as it was created.

Hutton never veered from his central argument. Earth was a machine, he declared from the beginning, designed by a divine mind for a particular purpose: 'The globe of this Earth is evidently made for man. He alone, of all the beings which have life upon this body, enjoys the whole and every part ... and he alone can make the knowledge of this system a source of pleasure and the means of happiness.' To us it is a magnificent, alien, hubristic vision: of Earth as a machine designed by God, on which volcanoes are nothing but valves to get rid of excess heat – a machine designed to sustain life and specifically mankind, like a huge space station.

But how old was the Earth? Hutton could not say how many cycles might have preceded the present one. In the relentless churning, all vestiges of the deepest past had been erased, and the trajectories of the far future could not be discerned. As Playfair would write, 'The author of nature ... has not permitted in his works any symptom of infancy or of old age, or any sign by which we may estimate either their future or their past duration.' Hutton was *not* saying the Earth was eternal (a point that was to be greatly misunderstood, as will be shown later). But if there was no trace in the present condition of the Earth of its origin or end, there was nothing he could say about those singularities: they were beyond the scope of 'human observation'.

It was a stunning declaration. Suddenly the walls of Ussher's box-universe had collapsed, to be replaced by a vast hall of mirrors, in which there was nothing to be seen but geological cycles, repeating over and over, each cycle enormously long – and the whole sequence unimaginably longer still. As Hutton thrillingly declared, 'The result, therefore, of our present enquiry is, that we find no vestige of a beginning, no prospect of an end.'

'We have now got to the end of our reasoning ... We have the satisfaction to find, that in nature there is wisdom, system, and consistency.'

Considering Hutton's presentation with the benefit of hindsight, we can recognise startling insights – in the heat origin of igneous rocks like basalt, in the 'rock cycle' (as modern geologists call Hutton's cycles of erosion and uplift), in Hutton's uniformitarianism, and in his intuition over the scale of Earth's true age.

There are also errors and omissions, however. Hutton was wrong to claim an igneous origin for flints, although ironically their presence in continental rocks was one of the first geological puzzles to attract his attention. More fundamentally, while Hutton rejected the old Biblical timescale for the creation of the world, he was prepared to concede, wrongly, that *human* history might match Ussher's narrow timescale: 'Now, if we are to take the written history of man for the rule by which we should judge of the time when the species first began, that period would be but little removed from the present state of things. The Mosaic history places this beginning of man at no great distance; and there has not been found, in natural history, any document by which a high antiquity might be attributed to the human race [in Hutton's day no prehistoric human fossils had been identified]. But this is not the case with regard to the inferior species of animals, particularly those which inhabit the ocean and its shores. We find in natural history monuments which prove that those animals had long existed; and we thus procure a measure for the computation of a period of time extremely remote, though far from being precisely ascertained.'

Hutton also believed that his previous ages were essentially similar to the present, save only for the peculiar absence of man; his study of the rocks had not demonstrated to him that previous ages, populated by dinosaurs or even lifeless, might be different to the present. It appears that Hutton's thinking here was shaped by his background. He had learned his geologising in Scotland, where 'the strata ... being much broken and confused, it is seldom that any one bed can far be traced'. There was nothing comparable to the reasonably regular upward succession of fossil-bearing rocks to be found in southern England, which would give the first geological map-makers their clue in their mapping of geological ages with their cargoes of evolving life.

Besides, Hutton's interest in the rocks had always been mineralogical and chemical. He had never been much concerned with any fossils he observed: 'Being neither botanist nor zoologist in particular, I never considered the different types of figured bodies found in a strata, further than to distinguish betwixt animal and vegetable, sea & land objects.' Indeed, his superficial study of the fossils he did observe misled him: 'We have but to examine the strata of our Earth, in which we find the remains of animals. In this examination, we not only discover every genus of animal which at present exists in the sea, but probably every species, and perhaps some species with which at present we are not acquainted. There are, indeed, varieties in those species, compared with the present animals which we examine, but no greater varieties than may perhaps be found among the same species in the different quarters of the globe. Therefore, the system of animal life, which had been maintained in the ancient sea, *had not been different from that which now subsists*, and of which it belongs to naturalists to know the history' (my emphasis). Of course he was wrong; the animals of antique times *were* different from the present.

Hutton missed the possibility, then, of recognising a progression of life in the fossils, as well as of using them as an index to date strata; educated by Scotland's broken rocks, he was always more struck by the great violence done to the land than by its orderly construction.

Moreover, his insights are set in an uncertain matrix of ill-formed physical theory. Here Hutton was a victim of his times: physics and chemistry were simply not mature enough to provide a proper context for his ideas, and his appeal to divine design arguments would certainly not be scientifically respectable today.

At bottom, however, his hypothesising was based on what he had found out in the field, and the instincts he had developed on his windswept farms; and his intuitive grasp of Earth's cyclic unity was profound and true. It was certainly a pivotal moment in scientific history. Hutton had been the first to set out a coherent and testable modern model of the Earth to compete with the creaking Neptunist notions – and in the process he had discovered deep time.

After the completion of the reading of the second part of his paper, to the applause of his colleagues Hutton sat down. The whole of his argument is an elegant interplay of three key metaphors: the Earth as an orderly Newtonian system, as orderly as the heavens; the Earth as a machine, like Watt's steam engines; and the Earth as a body with cycles of renewal, like Harvey's circulating blood. At last all the threads of Hutton's life had come together: the disciple of Newton and Maclaurin, the doctor, the farmer, the visionary geologist. It was the high point of his intellectual life: he must have felt that he had achieved his ambition of becoming the Newton of the Earth.

But this was Edinburgh's Royal Society, not a gentlemen's club. And Hutton's audience, in the lecture room and beyond, would be harsher in its reception of his ideas than he could have imagined.

15

'The world was tired out with geological theories'

On 4 July 1785, three months after Hutton's oral presentation, a meeting was called to discuss his paper. Hutton was evidently nervous, for he was again struck by illness. And well he might have been: clubbability and charm would be no use when the hard questioning started.

Every graduate student of science goes through a similar process. The conclusion of my own doctorate (in engineering) was a 'viva', an oral presentation of my work, with tough scrutiny by experts from my own department and from another university. Such sessions are gruelling, and, scientists being as human as everybody else, there is always plenty of backbiting and score-settling. But the purpose is to make the science better, by driving out errors and misapprehensions. In a way, modern science works by harnessing some of the less pleasant aspects of human nature as a feedback mechanism to improve the work itself.

James Hutton was not a professional academic, however. He was fifty-nine years old, and the medical degree he had obtained as a young man was now a distant memory. Not only that: he also found – perhaps to his surprise – that much of his audience, even in the Royal Society of Edinburgh he had just helped set up, was indifferent, even hostile.

Hutton's presentation itself had not impressed. He was not a lecturer. Though it did contain striking phrases – 'no vestige of a beginning' – and though he admirably eschewed excessive technical jargon, his paper was awkwardly written, and difficult to follow. Even his key supporter Playfair would say that 'the reasoning is sometimes embarrassed by the care taken to render it strictly logical; and the transitions, from the author's peculiar notions of arrangement, are often unexpected and abrupt. These defects run more or less through all Dr Hutton's writings, and produce a degree of obscurity astonishing to those who knew him, and who heard him every day converse with no less clearness and precision, than animation and force.' This was a major strike against Hutton in a time when science was seen almost as a branch of literature – Playfair called Hutton the 'author' of his theory – and the elegance with which an argument was presented was a significant factor in its authority.

Besides, though chemistry and mineralogy were frequently discussed topics within the Society, geology was not. It may have been difficult for the less geologically attuned in Hutton's audience to grasp the points he was making about his specimens. Playfair would say that Hutton's descriptions 'suppose in the reader too great a knowledge of the things described'.

Even for those who understood his geology, Hutton's theory was off-puttingly radical. His views on granite and basalt, perhaps the best founded on his field observations, were diametrically opposed to prevailing geological theories: by placing granite among the youngest rocks on Earth's surface, he completely reversed the time sequence of Werner and his Neptunists. The big leap was difficult to accept. In addition, nobody had really understood Hutton's careful epistemology and his uniformitarianism, or his arguments about heat – partly because he hadn't sufficiently explained them, in a presentation Playfair called too brief.

Then there was the man himself. Popular he might have been, but even his closest friends were prepared to admit that Hutton had long been something of an oddity in Edinburgh's fashionable circles: the farmer without a formal academic position, with antiquated views on phlogiston and the like, with his eccentricities of manner and dour clothes. Given the grandiose nature of his theory, a certain snobbery may have cut in among the assembled academics.

Then there was the evidence – or the lack of it. Hutton could have called on results from geological experiments carried out by Black, himself and others on what happened when you melted and cooled rocks in the laboratory. But oddly, he hadn't. By this time Hutton was discouraged about the value of experimentation; he would argue with Sir James Hall about it. Surely no human experiment could yet match the pressures and temperatures that were possible in nature? Perhaps Hutton feared that quoting an imperfect experiment might do his theory more harm than good.

Even Playfair conceded this had been a mistake. Though Hutton himself believed the evidence he showed for the action of heat on the rocks was very strong, 'for my part,' wrote Playfair, 'it is a conviction that would be strengthened by an agreement with the results even of such experiments as it is within our reach to make. It seems to me, that it is with this principle in geology, much as it is with the parallax of the Earth's orbit in astronomy; the discovery of which, though not necessary to prove the truth of the Copernican System, would be a most pleasing and beautiful addition to the evidence by which it is supported.' With insufficient evidence, Hutton looked, despite all his epistemology, like nothing but another old-fashioned geological theorist: even Playfair would say, 'The world was tired out with unsuccessful attempts to form geological theories ...'

John Walker had listened with particular hostility to Hutton's presentation. Walker, professor of natural history at the University of Edinburgh from 1779 and nemesis of the Scottish antiquarians, was now the secretary of the Society's Physical section, to which Hutton presented his theory. Walker had known Hutton since 1770, but, as Hutton surely knew, Walker was an open opponent of geological theorising of any kind. He was also an influential teacher whose students had included Sir James Hall and Playfair – and a young man called Robert Jameson, who would later give Hutton's theory and his legacy a very tough time indeed.

Perhaps what hurt Hutton most of all, though, was the indifference with which his presentation was received. Some of his audience did get the point: 'His paper contains a variety of ingenious observations and new facts established principally from specimens in his own instructive geological collection,' one would write. But as Playfair wrote, you might have thought that 'a work of so much originality as this Theory of the Earth ... would have produced a sudden and visible effect ... yet the truth is .... that several years elapsed before anyone shewed himself publicly concerned about it, either as an enemy or a friend.'

And then there was also the question of God. There were plenty of Presbyterians in the audience – both Walker and Playfair were ministers of the Kirk – and any conventional Christian would recognise that Hutton's 'no vestige, no prospect' was a clear rejection of the Genesis narrative. What made it worse was that Hutton's epistemology and his notions of uniformly working natural processes clearly echoed the writings of David Hume, the notorious God-denier. James Hutton, it was said, had shown himself to be an atheist.

Of course, the charge was misdirected. Hutton's design-argument theory was nothing if not a vision of how God operates. Indeed, it seems that a key impulse for constructing his theory was a search for a new faith after his life on the farm had left 'nothing of the Christian about me'. Of the attacks he would endure in the years to come, none would upset him more.

His very first reaction after the Society session was to try to respond to the religious questions. In that same month of July he wrote out a new preface to his work, entitled 'Memorial Justifying the Present Theory of the Earth from the Suspicion of Impiety'. He tried to argue that religion and science should not come into conflict because they held authority in different spheres. The Book of Genesis was not a literal diary of Earth's creation but a kind of celebration of God's power. In any event, he insisted, he was not being impious: 'The Word of God, whether revealed by the common faculties of man or given to human understanding in a preternatural manner, must be always one.'

He sent his draft to his old friend William Robertson, by then the principal of Edinburgh University. Robertson famously opposed religious bigotry, and his daughter Mary had actually married a prominent Deist. Robertson smoothly rewrote Hutton's preface, but he did wonder whether Hutton should bother with a preface at all. After all, it wasn't what he truly believed. Hutton and Robertson concluded that the preface would do more harm than good, and Hutton quietly dropped it.

A pamphlet summarising Hutton's thesis was produced shortly after his talk. It spanned twenty-eight pages and was entitled 'Abstract of a Dissertation Read in the Royal Society of Edinburgh, Upon the Seventh of March, and Fourth of April, 1785, Concerning the System of the Earth, Its Duration, and Stability'. It is the first known publication of Hutton's theory. Few copies of the pamphlet were produced, and fewer still survive; it was seen as ephemeral, and copies would have been destroyed when the full version of Hutton's paper eventually appeared in the *Transactions*. Two copies came on the market when Sir James Hall's library was sold off in 1947. They went for £90 and £110 – still in their original blue paper wrappers, and signed 'From the Author Dr Hutton'.

The Abstract was circulated in Britain and on the Continent. In France it was read by Desmarest, for instance. Hutton received an enthusiastic response from one Matthew Guthrie, a founder of the Royal Society of Edinburgh and Physician to the Imperial Cadet Corps in St Petersburg, who replied, 'I am much flattered by your attention in sending me a copy of your little publication. I am only sorry that I cannot satisfy a desire it naturally creates, that of seeing your dissertation at large ... Your mode of investigation and choice of proofs from natural phenomena which stare every man in the face, and from matter that chemistry can demonstrate the changes it has undergone, certainly appears to me as the most probable manner of coming at truth ... I wish you therefore much success and much honour which will naturally result from it.'

Hutton knew, however, that the Abstract wasn't extensive enough to provoke a wide response, positive or negative. For that he would have to wait until the formal publication of his work in the Society's *Transactions*, which was some years away. What was he to do in the mean time? He would get nowhere by picking away at the religious aspects. He had also been stung by the accusations that he hadn't produced enough evidence to back up his assertions.

It was time to consult God's books once more. To his credit – at the age of fifty-nine – Hutton went back out on the road, once again risking weary feet and saddle sores, to gather more data.

His first concern was granite.

A key premise of his theory was that granite was not a

'primitive' rock, the first to be laid down, as the Neptunists believed. To Hutton granite was a *young* rock, intruded in a molten state into older formations. The reason granite was predominant among Alpine peaks, for example, was because it was durable, not because it was ancient.

Only one of these granitic theories could be true. 'Dr Hutton', wrote Playfair, 'was anxious that an *instantia crucis* might subject his theory to the severest test.' This language shows Hutton was following classical scientific method. An *instantia crucis* – a 'crucial instance', a term defined by Francis Bacon – means a specimen or test which distinguishes unambiguously between two competing ideas. And now Hutton wanted to find a crucial instance regarding the nature of granite.

His mind turned to the Scottish Highlands. He knew that there was granite at the source of the River Dee, and schist at the source of the Tay (schists are stratified rocks of various kinds, with a high content of mica or hornblende). So the countryside between the two rivers looked promising for a junction between the two types of rock. John Clerk of Eldin, in fact, had already seen granite veins in the River Garry. What Hutton hoped to find were examples of granite pushing through sedimentary rock, which would prove the granite was younger than the sedimentary, and that it must have flowed in a molten form.

So in September 1785 – just a few months after the unsatisfactory presentation to the Royal Society – Hutton and Clerk set off to stay with the fourth Duke of Atholl, whose deer forest, close to the Duke's seat of Blair Castle, contained the sites of interest.

This noble family had been split during the rise of 1745. The incumbent Duke had supported the government, but his brothers favoured Prince Charles. The Duke eventually fled, and the Jacobite youngest brother laid siege to the castle. (Blair would be the last castle ever besieged in mainland Britain.) The successful Jacobite brother took the estate and went on to father the third Duke, who in turn fathered the fourth – so the young man who welcomed Hutton and Clerk on their geologising trip was the grandson of the Lieutenant-General of Bonnie Prince Charlie.

Hutton and Clerk made their first base in the Forest Lodge,

about ten kilometres up the valley of the Garry. It is a beautiful area: Burns would stay there two years later, as would Wordsworth and his sister in 1803. Clerk and Hutton had arrived at Blair in the shooting season, and the expedition, Hutton recorded, 'made an agreeable party of pleasure of a thing which otherwise would have been incommodious and painful'.

In the company of the Duke, Hutton and Clerk travelled along Glen Tilt, a tributary of the Garry. The Glen is a long, narrow valley where the river has exposed the underlying structure of the rock. They had some trouble: they had to climb a narrow rivulet, a route which was 'fit for no other than the footsteps of a goat'. Clerk made drawings of the area for his friend, and Hutton would later make a beautiful watercolour map (which is now in a US Geological Survey collection).

Perhaps, as they climbed, they talked about the great affairs of the world.

In recent years John Clerk of Eldin – once a medical student, mine operator and etcher – had been developing yet another new career. He had been intrigued by accounts of an abortive naval battle between England and France in 1778, during the American War of Independence. The debacle had led to the British commander-in-chief, Admiral Keppel, being court-martialled for misconduct and neglect of duty. During 1779 the case, which lasted five weeks, was discussed with great interest. Keppel was eventually acquitted – but no naval battle had ever been documented so thoroughly.

As he read these reports, Clerk began to muse on principles of naval warfare, so vital to the ongoing war with France. Battleships of the time fired their guns in broadsides, so opposing fleets would be drawn up in lines of battle, sailing past each other on parallel courses. Clerk wondered if it might be more effective to try to break the enemy's line, which would throw his ships into confusion. He would demonstrate his ideas with little cork models on Edinburgh dinner tables. The young Walter Scott would pocket some of the models, with Clerk complaining good-humouredly that his demonstration had been ruined. Perhaps predictably, Clerk faced great frustration trying to get anyone in authority to listen to him. But his ideas had been applied at last in a great victory over the

French fleet in the Windward Isles, to much celebration in the Edinburgh circle.

At Glen Tilt, Hutton and Clerk found exactly what they were looking for: 'granite breaking and displacing the strata in every conceivable manner' – proof that the granite was indeed younger than the strata into which it had intruded. Hutton was so noisily delighted that 'the guides who accompanied him were convinced that it must be nothing less than the discovery of a vein of silver or gold, that could call forth such strong marks of joy and exultation'.

But the trip wasn't done yet. Braving discomfort – 'in matters of science, curiosity gratified begets not indolence, but new desires' – the friends abandoned the Lodge and penetrated further into the wilderness of the Highlands, until they reached a still more remote hunting seat at Fealar, 'the most removed, I believe,' Hutton would write, 'of any in Britain from the habitations of men'.

While the Duke went shooting – he bagged three harts and a hind, 'all in excellent condition' – Hutton and Clerk walked up the valley of the Tarf, a tributary of the Tilt. They found more treasures in tumbled fragments of schist with granite intrusions. In one sample they found a vein of granite that pushed through an older mass of granite as well as broken schistus. Here was proof of a whole series of geological events: the sediment had been laid down, the first bit of granite had intruded, and then after that another vein of granite had pushed through the composite mass.

From Glen Tilt Hutton brought home a *boulder* of granite, intending to show its igneous origin. The difficulty of transporting such a monster across hundreds of kilometres by sailing ships and horses is a measure of the importance of Hutton's samples to him. When Hutton wrote up this excursion he would conclude his Glen Tilt observations by comparing the geology of the Lowlands and Highlands, and remark, with underlined emphasis, '*That whatever be the materials in those two cases, Nature acts upon the same principle in her operations, in consolidating bodies by means of heat and fusion, and by moving great masses of fluid matter in the bowels of the Earth.*'

*

The following September, Hutton and Clerk set off again, hoping to investigate granite and other formations on the western island of Arran, but the weather was too poor for them to make the crossing. They explored the Clyde coast instead, travelling from Glasgow through the shires of Ayr and Galloway. They found basalt dykes and granite exposures, and Hutton observed that the Rinns of Galloway, now a peninsula, had once been an island – proof that the sea level had been higher in the past. In Galloway, granite had become prized as a building stone, and Hutton quizzed the locals who had become expert on its nature. They also visited a lead mine in which Clerk had an interest, and consulted its overseer for local geological information.

Hutton, now sixty, remained vigorous. One day, near Sandyhills Bay on the Solway, the geologists had been walking beside their chaise along a road. But when the road turned inland they gave up the chaise and scrambled along the sandy bay to a rocky portion of the shore. Here the strata in the schist rock were tilted almost upright, but granite broke through the sandy beach. They looked for the junction between the granite and the schist, but it was covered in bushes and briars, through which the two elderly explorers had to push their way. At last they found a neat granite vein intruding into the schist, just as they would have hoped, dwindling to a thread where it could penetrate no further.

Another day, the geologists struggled to make out formations on the south side of the Cairnsmore mountain. Hutton ruefully remarked, 'To a naturalist nothing is indifferent; the humble moss that creeps upon the stone is equally interesting as the lofty pine which so beautifully adorns the valley or the mountain: but to a naturalist who is reading in the face of rocks the annals of a former world, the mossy covering which obstructs his view, and renders indistinguishable the different species of stone, is no less a serious subject of regret.'

By the end of this trip, and with the evidence from Glen Tilt, Hutton would write, 'We may now conclude, that, without seeing granite actually in a fluid state, we have every demonstration possible of this fact; that is to say, of granite having been forced to flow, in a state of fusion, among strata broken by a subterraneous force, and distorted in every manner and degree.' Back home he would still have to make this case – but Hutton, in this Indian summer of his geologising, seemed to have been rejuvenated.

As Hutton sought out his crucial instances, the debate that was shaping the future of geology continued.

In 1787, the arch-Neptunist Werner published a little book entitled *Short Classification and Description of the Rocks*. The 'Kurze Klassifikation', as it came to be known, was an important work in that it sent out a new and functional vocabulary and definition system that would help enormously in the classification of rocks in the years ahead.

But Werner also used it to reassert unequivocally his view that basalt has an aqueous origin. By now this claim was controversial even outside Hutton's circle, because field workers exploring the extinct volcanoes in Auvergne, France, had gradually reached the conclusion that basalt was much more likely to be igneous. Nonetheless, these weighty pronouncements from the leading figure in the field, utterly contradictory to Hutton's own ideas, must have deepened Hutton's unease.

Meanwhile, Edinburgh life continued. Robert Burns visited the city from 1786 to 1788: while there, he would write one of his most famous love poems, addressed to the niece of Colin Maclaurin. In 1787 the publisher William Creech threw a party to celebrate his Edinburgh edition of Burns's poems. It was at this party that Burns, aged twenty-eight, met Walter Scott, then sixteen, for the only time. Hutton and Black were both there – as was a glamorous aeronaut called Vicenzo Lunardi.

French balloonists had been performing spectacular aerial stunts since 1783. There had been a few flights in Scotland before, but Lunardi, secretary to the Neapolitan embassy in London, was a sensation. He made a spectacular take-off from the grounds of Heriot's Hospital in Edinburgh, and the sight of the immense object drifting in the sky, with handsome Lunardi in his Neapolitan army officer's uniform beneath it, caused consternation and excitement. Lunardi was rewarded with fame: he was wined and dined by Edinburgh society, and given honorary membership of the Beggar's Benison.

*

In August 1787, Hutton tried to reach Arran once again.

Hutton and Clerk had engaged a Dr Irvine to be their 'conductor in taking a vessel down the Clyde and visiting several places by the way', but he had unfortunately died in the spring. In August the weather was very poor, and Clerk of Eldin didn't want to undertake the journey. Hutton considered travelling alone, but to his pleasure Clerk's son, another John Clerk, offered to accompany him. The younger Clerk had a contracted leg, which made him limp, but he gamely pursued Hutton as they climbed Goat Fell, the highest mountain on the island – just to see a few basalt dykes, 'an idea which could not have entered the head of any sober person who was not a mineralist'. This time it would be the younger Clerk's job to make the geological drawings, which he did, expertly.

The younger Clerk, then thirty, would be called to the Bar in 1785. He found fame as an advocate in the notorious trial of Duncan Brodie. A member of the Town Council by day and a burglar by night, Brodie's double life would inspire Robert Louis Stevenson to write *Dr Jekyll and Mr Hyde*. Clerk was no teetotaller. Once, plastered early in the morning, he stopped a servant-girl in the street and asked where John Clerk's house was. 'Why, you're John Clerk,' she said. 'Yes, but it's his house I want.' Throughout his life he would retain his strong Scots accent. One day, addressing the House of Lords, Clerk argued that 'the watter had rin that way for forty years'. The Lord Chancellor sarcastically asked if the Scots spelled water with two t's. 'No, my Lord,' said Clerk. 'We dinna spell watter wi' twa t's, but we spell manners with two n's.' No doubt Clerk was a fine companion for Hutton on this summer jaunt.

Hutton would write of the purpose of this Arran expedition, 'It may now be observed, that the present history with regard to the island of Arran, is not proposed as necessary in that question concerning primitive mountains, nor as adding any new light to the nature of granite, as already investigated; but as an example in Cosmogony, where it may be proper to see the connection of various things, or the several parts which enter the constitution of the globe.' That is, by this time he was confident enough to proceed from particular bits of evidence to generalities, and to see 'how far the natural history of Arran shall be proper to try the Theory of the Earth which had been formed from that of other parts'. This small Scottish island was to serve as a scale model of a planet.

Again Hutton investigated junctions between granites and schists. Once they had caught the scent of a contact, the two men followed their motionless quarry 'with more animation than could have been expected from such an innocent chase'. Hutton was startled by the number and complexity of the dykes of basalt he found, and in one place he found dykes made of black glass, which pleased him greatly, for they were overwhelming evidence of an igneous origin for the basalt.

The island, crammed with geological phenomena, proved so interesting to Hutton that he planned to reconstruct its geological history, but unfortunately this was a project he never carried out.

Hutton still needed to find clinching proof of his suggestion that the land was shaped by his cycles of erosion, consolidation, uplift. A 'crucial instance' in this case would be a place where the remnants of strata from one previous cycle had been lifted, broken, eroded, and then overlaid by strata from the next cycle: an 'unconformity', in modern geologists' terms.

He had actually seen one such example at Lochranza in Arran. Having set out on horseback to explore the feature, he found that the River Sanox ran along the junction between the schist and the granite of the mountains. So he abandoned his horse and scrambled over the moss and the moor to the head of the river. He found his unconformity, but it wasn't as convincing an example as he would have liked – Playfair wrote that 'the contact ... is so covered by the soil, as to be visible in very few places'. Nonetheless, Hutton's discovery is now lauded by historians of geology, for it was the very first unconformity of its type to be recognised in Britain.

On his return from Arran in 1787, Hutton happened to visit a friend in Jedburgh, in the border country of south-east Scotland. And there he found, purely by chance, a remarkably clear example of what he sought.

Jedburgh was a quiet, beautiful place. Its abbey, founded in the twelfth century, repeatedly found itself in the way of marauding English armies; it was burned three times in the fifteenth century alone, and again by Henry VIII's armies in

1544. The Union of Scotland and England brought Jedburgh relief from the armies, but little economic benefit, even by 1787 – the town was a poor place. But history runs deep: today, in the roofless ruins of the abbey, you can see Roman altar stones, cut up and reused by medieval masons.

During his visit, Hutton took a walk south along the bank of the River Jed. The valley is narrow but deep, its walls mostly clad in green. In one place, however, erosion had exposed the underlying rock. And Hutton was startled to see, as clear as day, an unconformity of just the type he had been seeking.

You can visit the exposure today. It is known as 'Hutton's unconformity', and even the local tourist office knows about it – though its literature sites it in the wrong place. The exposure is on private land, and you have to phone ahead to the owner, a kindly gentleman called Mr Veitch. There is a path, mostly overgrown, leading down from the verge of the A68, the main road to Newcastle and Edinburgh. It's something of a scramble: Mr Veitch told me that erosion regularly makes it impassable.

The bedrock here is sandstone and marl pierced by veins of basalt. You can see fragments of these veins sticking up out of the river itself, a feature Mr Veitch, in his childhood, learned to call 'clints'. The exposure itself is small, a purple-red gash in the river bank foliage only a few metres high and several paces long. But in this green-framed window you can see two types of strata. Playfair describes how 'the schistus there is micaceous, of vertical plates, running from east to west, though somewhat undulated. Over these is extended a body of red sandstone, in beds nearly horizontal, having interposed between it and the vertical strata a breccia full of fragments of these last.' The strata of underlying greywacke, a Silurian-age conglomerate of quartz and slate, are tilted vertically. These strata are very fine and sharp, some no more than paper-thin. But overlying the vertical striping of the greywacke, the horizontal bands of red sandstone are easy to see.

This contact of vertical with horizontal is the key to Hutton's reading of the feature. The two sets of strata, horizontal and vertical, must have been laid down at different times and in different conditions; the rocks can only be interpreted as the ruin of one cycle laid down unconformably over another.

John Clerk made a charming copper-plate of the unconformity. The picture (reproduced on the jacket of this book) shows a phaeton and horseman meeting on flat and unremarkable dry land that lies over what had been marine strata, violently distorted. My own photos, taken in 2002, show a similar cross-section, with modern four-wheel-drives running along the A68 above the same strata. Hutton's unconformity is a small feature, unnoticed by the traffic that thunders above, but if you can read the rocks, it tells a dramatic story.

Mr Veitch gets few visitors. The town council has a plan to make the path properly and even build a viewing platform. But Mr Veitch says the same plans have been under discussion as long as he has lived there, which is fifty-five years – a quarter of the way back to Hutton himself. In the mean time he has to call in the fire brigade regularly to hose away the vegetation which forever threatens to overwhelm the unconformity.

Still, Jedburgh is proud of its bit of geological history. On plaques around the town you will see Hutton's portrait displayed as one of four 'Famous Folk Linked to Jedburgh' – along with one of the few prominent female scientists of the nineteenth century, the author of 'Rule Britannia', and the inventor of the kaleidoscope.

After the serendipity of Jedburgh, Hutton now deliberately sought more examples of well-exposed unconformities.

He knew that the boundary between the resistant greywackes of the Southern Uplands and the softer sandstone of the low country ran through the Dunglass estate, some sixty-five kilometres east of Edinburgh, home of his old friends the Hall family. At Hutton's request, Sir James Hall had his uncle search for the contact between greywackes and sandstone in a burn called the Tour. Then, in June 1788, Hutton and Playfair joined Hall at Dunglass.

Playfair, Hall and Hutton took a boat and set off to explore the coast. The weather was calm enough for them to be able to steer close to the foot of the rocks. They followed the sandstone as it rose towards the schist.

At last they came to the headland called Siccar Point. Here the junction between the rock types had been washed bare of vegetation by the sea. When they landed they could see clearly how, just as at Jedburgh, the older strata had been tilted on end and eroded before younger sediments, made up of debris from the older, had been laid flat on top. Playfair's record of his response to this, in his biography of Hutton, remains a classic of science writing:

> What clearer evidence could we have had of the different formation of these rocks, and of the long interval which separated their formation, had we actually seen them emerging from the bosom of the deep? We felt ourselves necessarily carried back to the time when the schistus on which we stood was yet at the bottom of the sea, and when the sandstone before us was only beginning to be deposited, in the shape of sand or mud, from the waters of a superincumbent ocean. An epocha still more remote presented itself, when even the most ancient of these rocks, instead of standing upright in vertical beds, lay in horizontal planes at the bottom of the sea, and was not yet disturbed by that immeasurable force which has burst asunder the solid pavement of the globe. Revolutions still more remote appeared in the distance of this extraordinary perspective. The mind seemed to grow giddy by looking so far into the abyss of time; and while we listened with earnestness and admiration to the philosopher who was now unfolding to us the order and series of these wonderful events, we became sensible how much farther reason may sometimes go than imagination can venture to follow.

Hutton had at last taught himself to read God's books, and the stories he could tell were wonderful indeed.

Hutton and Playfair walked back along the burn, looking for more junctions between the strata. They failed to find any, although they did see basaltic boulders from a large dyke that cut through the burn. Later in the day the rain came down, washing out a further expedition they had planned. Still, it had been a good day. They had 'collected, in one day,' noted Playfair, 'more ample materials for future speculation, than have sometimes resulted from years of diligent and laborious research'.

Playfair's judgement was surely right. Hutton has been retrospectively criticised for publishing his theory *before* assembling crucial evidence such as the clinching granite samples, and he hadn't seen a single unconformity when he stood before the Royal Society of Edinburgh in 1785: any modern research student would be criticised for such lapses. But even before 1785, Hutton had clearly understood the rocks on which his theory was founded. On his return to the field, he knew exactly what he was looking for and how to find it, and the careful, intelligent, directed forays of these later years bore ample fruit. To my mind, there is no doubt he was a good scientist, a good geologist.

Hutton made more trips that summer, including a visit to the Isle of Man with the Duke of Atholl. The Duke, who owned the island, had asked Hutton and Black to make a mineralogical survey of the island. Though he enjoyed the Duke's hospitality, Hutton found nothing there that modified his theories. On the way back home, he visited the Lake District. In a quarry near Windermere he found a piece of limestone with fossil impressions in it – neat proof that another supposedly 'primitive' rock was no such thing.

But these weren't major or purposeful expeditions. On that memorable day at Siccar Point, none of the companions could have known that this was the last significant geological exploration James Hutton would make.

16

'A wild and unnatural notion'

At last, in 1788, three years after his oral presentation, Hutton's paper appeared in the *Transactions of the Royal Sociey of Edinburgh* (although preprints may have been circulating for a year or so before that). I studied the paper in the library of the Royal Society of Edinburgh, in the Society's bound set of its *Transactions*. The paper has survived the centuries well, and the somewhat archaic typeface is readable, but the plates, drawings of some of Hutton's hand samples, have stained the facing pages.

The phoney war was over. With proper publication, Hutton's ideas were finally exposed to a national and international audience. He waited nervously for the reviews to appear.

The reaction was mixed, and disappointing.

The *Analytical Review* dismissed the theory in a paragraph as just another 'philosophical romance'. The *Critical Review* gave Hutton four pages. Though the reviewers wouldn't quite back his theory, the empirical evidence was discussed, and the majesty of his thinking praised: 'the mind cannot comprehend so vast a system'.

The *Monthly Review*, though, doubted that subterranean fire consolidated the rocks, and was shocked by what they saw as Hutton's notion of 'a regular succession of Earths from all eternity!' Hutton, of course, had not claimed an eternal age for the Earth, just that the beginning of the Earth, like its end, was beyond the scope of human investigation – a subtle difference, but crucial from a theological point of view, and endlessly misunderstood.

Hutton got some support from his old friend Erasmus Darwin, who used Hutton's theory to inform his *Botanic Gardens*, a poem on botany and geology. Hutton and Darwin disagreed about geology, but Darwin's poetry was extremely popular, and his work won Hutton some welcome publicity.

The first major riposte to Hutton's theory was much more hostile. It was a forty-page attack made in 1789 by John Williams in his book *The Natural History of the Mineral Kingdom*. Williams assailed all Hutton's major points, but focused his fire once again on his supposed claim of an eternal world. This was a 'wild and unnatural notion' that led to 'scepticism, and at last to downright infidelity and atheism'.

If Hutton was uneasy about the reception of his theory, then he was more confident about his results on granite. In 1790, therefore, he decided to present the results of his explorations of Glen Tilt, Galloway and Arran to the Royal Society of Edinburgh, and show the specimens he had brought back. He was shaken by the response even to this. Hutton hadn't done enough, his Neptunist opponents said, either in his specimens or in his accounts of his field trips, to establish his case that granite was young and heat-moulded, rather than ancient and water-deposited.

In 1791, Werner weighed in again from his Saxony fortress. This time he published a theory of mineral veins. Such veins had nothing to do with heat, as Hutton claimed; according to Werner they were all aqueous in origin – even basalt – and they had all been injected into the strata from *above*, seeping down from oceans into cracks in the rock.

Hutton also came under attack in 1790–91 from Jean André Deluc in *The Monthly Review*. Deluc, a weighty figure, was Swiss-born but resident in Britain. His greatest legacy would be his work on meteorology, a subject over which he and Hutton had already fallen out. In 1788 Hutton had published a 'theory of rain', based on the effects of heat on humidity, drawn from decades of observations of the weather. Deluc's attack on this was vigorous, and 'the controversy was carried on with more sharpness, on both sides, than a theory in meteorology might have been expected to call forth', as Playfair dryly noted. As a geologist, Deluc was an old-fashioned thinker; his keenest interest was in reconciling the rocks with the Genesis story. Now, over seventy pages, he laid into Hutton. He refuted Hutton's evidence of erosion and consolidation, and vowed to oppose his dismissal of miracles and other supernatural causes. It was the most formidable charge of atheism yet brought against Hutton. Playfair believed that Hutton drafted a reply to Deluc which the editor of the *Review* refused to insert, but Playfair hesitated to bring a 'positive charge of partiality against men who exercise a profession in which impartiality is the first requisite'.

There were aspects of these various reactions that Hutton probably welcomed. He was not subject to personal attack, and at least he was taken seriously, as the number of long and weighty reviews must have told him. Even so, the reaction to his theory remained scattered and muted, and the continuing attacks over his supposed atheism were troubling. And by now, time, though limitless for the Earth, was running out for Hutton's circle.

Adam Smith had enjoined his friends that they should destroy his lecture notes after his death, but permitted them to use the rest of his manuscripts as they pleased. 'When now [Smith] had become weak,' Hutton wrote in the *Transactions of the Royal Society of Edinburgh*, 'and saw the approaching end of his life, he spoke to his friends again upon the same subject. They entreated him to make his mind easy, as he might depend upon their fulfilling his desire. He was then satisfied. However, some days afterwards, finding his anxiety not entirely removed, he begged one of them to destroy the volumes immediately. This accordingly was done; and his mind was so much relieved, that he was able to receive his friends in the evening [of one Sunday in 1790 at Smith's last home, Panmure House] with his usual complacency.'

Smith's friends had, Hutton continued, 'been in use to sup with him every Sunday; and that evening there was a pretty numerous meeting of them. Mr Smith not finding himself able to sit up with them as usual, retired to bed before supper; and, as he went away, took leave of his friends by saying, "I believe we must adjourn this meeting to some other place." He died a very few days afterwards.' Smith was buried in the Canongate, Hutton and Black serving as his executors.

And then – in 1791, aged sixty-six – Hutton himself fell ill.

Until this point in his life he had been lucky enough to enjoy good health. But now, as Playfair reported, 'The disorder that threatened him (retention of urine), was one of those that most immediately threaten life, and he was preserved only by submitting to a dangerous, and painful operation.' The operation was performed by Black and his colleagues – without anaesthetic. Hutton was left weakened and stranded in his room. He would thereafter suffer from his ailment periodically; at times he was bedridden.

Because of the sparseness of Playfair's account, we cannot be sure what it was that afflicted Hutton. Given his age, it was most likely a prostate problem. If it had been cancer he would probably have been killed more quickly, but benign prostatic hypertrophy would have caused these symptoms and could well have recurred. Hutton's earlier life might have made venereal disease a possibility; a urethral stricture would have been a very painful cause of the retention of urine, but this would probably have affected him at an earlier age. In any event, it was probably the brutality of the surgical procedures in Hutton's day, and the poor quality of the aftercare (by modern standards) he would have received, that caused his subsequent deterioration.

For now, though, 'the goodness of his constitution, aided, no doubt, by the vigour and elasticity of his mind, restored him to a considerable measure of health, and rendered his recovery more complete than could have been expected'. Hutton would continue to write, read, receive samples from his friends, and welcome visitors. But it was the end of his geologising in the field. There would be no more hill-climbing, or deer-hunting dukes, or scrambles along rocky shores in search of intrusions and unconformities. The Earth had no more to show James Hutton.

By now, most European geologists were divided into two camps, neither of which had been influenced much by Hutton's theories.

The 'Vulcanists', including Desmarest and Faujas, were fire geologists, who believed that volcanoes must have had significant effects on the evolution of the Earth. The other school was the Neptunists, who thought volcanoes were irrelevant special effects. To them water was the key agent. All Neptunists hypothesised some kind of universal ocean, out of which the rocks had been deposited. But there was still a whole spectrum of theologically inclined thinkers, ranging from those who still held to the most literal interpretation of the Bible account, to others who interpreted its teaching in a more symbolic or allegorical way.

Of these various factions, Hutton had pretty much offended everybody. As a uniformitarian who relied on the working of present-day forces to shape the world, he made an enemy of anybody who believed in past catastrophes of some kind, such as the Neptunists. Not only that, he was advocating an age for the Earth that had to be much older than even the most generous estimates of the non-fundamentalist thinkers, like Werner himself.

Then there was the little matter of the Earth's internal heat. Hutton's thinking on subterranean heat earned him the label 'Plutonist'. But his position on this annoyed even the Vulcanists. There were problems if Hutton's theory was based on the then conventional ideas of caloric, the heat 'fluid'. Heat causes expansion, which was effected by caloric particles repelling each other. As long as you kept adding caloric, the expansion would continue, but if you stopped adding caloric, attractive forces would impose a contraction. So Hutton's heat engine would have to have a source that kept adding caloric *for ever* – which would surely cause an indefinite expansion of the Earth, which was absurd.

Another problem was the *source* of Hutton' s heat. None of the standard theories of the day seemed to offer any help. Could there be a central fire burning endlessly within the Earth? Perhaps, but where had all the phlogiston (or, if you followed Lavoisier, the oxygen) come from, to fuel such a blaze?

All this was based on a misunderstanding, as Hutton's own heat theory wasn't based on caloric – but then, he still hadn't said, even now, what it *was* based on. And he had made no claims about the source of the heat. The heat was just *there*, circulating without end, consolidating sediments and uplifting continents. This was unsatisfying for his supporters, and an open goal for his opponents.

Amid all this controversy, and largely confined to his home, Hutton continued to receive friends, to think, to read –

and to write. Hutton had always written prolifically: he used writing as much to clarify his own thoughts as to communicate his ideas to others. Now, even as his strength failed, he was trying to publish the results of a lifetime's contemplation.

Hutton's habit was to dictate his work. His secretary of this period had a graceful, flowing and very legible hand. Perhaps his patient amanuensis was his last surviving sister, Isabella. The women in Hutton's adult life, like the 'incendiary' Madam Young, are shadowy figures. Not so the mother who had raised him and ensured his education after the early death of his father, the sisters with whom he had lived most of his life, and especially Isabella, who now nursed him. But Isabella herself grew ill as the decade wore on.

Hutton's first book, published in 1792, was called *Dissertations on Different Subjects in Natural Philosophy*. Here he reprinted his paper on rain, with an appendix that reprised his bitter argument with Deluc. The rest of the book concerned his phlogiston theories, to which he was still much attached. He described how phlogistic matter would affect measurable properties of matter like hardness and ductility, and how phlogiston would affect light and colour. The book was touchingly dedicated to Black, acknowledging science's debt for 'your philosophical discoveries'.

No sooner were the *Dissertations* complete than Hutton began work on a mighty new project. This would be published in 1794 in three volumes, under the title of *An Investigation of the Principles of Knowledge, and of the Progress of Reason, from Sense to Science and Philosophy*. In this vast and baffling work, Hutton developed the epistemology he had built on Hume's philosophy: 'Body is not what it is conceived by us to be, a thing necessarily possessing volume, figure and impenetrability, but merely an assemblage of powers, that by their action produce in us the ideas of these external qualities ...' He set out his belief that what we perceive doesn't necessarily match anything that actually exists 'out there'. What we see is 'the creation of the mind itself, but of the mind acted on from without, and receiving information from some external power'. Playfair, characteristically, would put this rather better:

'External things are no more like the perceptions they give rise to, than wine is similar to intoxication, or opium to the delirium which it produces.'

It is an eerie idea, as if we live in a vast virtual-reality system, with sensations artificially injected into our minds. But Hutton went to great lengths to deduce the precepts of religions from this theory: because the universe as reconstructed by our minds is consistent and enduring, it is as real to us as if it *were* external reality, and so we have just as much an obligation to act morally.

Hutton's friends couldn't understand why he was spending his time on this turgid tome. They knew that he had always intended to expand his 1788 paper on the theory of the Earth into a longer book. He had been writing up his findings from his field trips and readings in a series of essays and papers, some presented to the Royal Society of Edinburgh, but he continued to put off the labour of completing the long geological work, and as his health remained fragile his friends began to fear that it might never appear in his lifetime – and that his theory would be left undefended.

Hutton knew what he was doing, however. His ragged epistemology was the context that contained the jewel of his geological methodology, his uniformitarianism.

Essential or not, Hutton's huge work was almost universally ignored by learned society, then and since. Even his closest friend, Black, wouldn't buy a word of it. Adam Ferguson joked that 'unreal as corporeal subjects were in his apprehension, he established a lucrative manufacture, on principles of chemistry.'

In 1793, Hutton's illness took a turn for the worse, leaving him further weakened. During his slow convalescence, he began to work through the proofs of his *Principles of Knowledge*.

And it was at this moment – with Hutton in his late sixties, frail and often bedridden, his theory barely making a ripple in the ocean of understanding – that he was subjected to the most withering attack of all.

Richard Kirwan, Fellow of the Royal Irish Academy, was a chemist, mineralogist, meteorologist and geologist. He was well known to Hutton through his defence of phlogiston. An eccentric recluse, he had a formidable mind, a deep faith, a profound sense of righteousness – and many chips on his shoulder. When he read Hutton's theory he was outraged.

17

'An abyss from which human reason recoils'

Kirwan was the second son of four. Seven years younger than Hutton, he was born in Cloughballymore, in County Galway, Ireland. The Kirwans were a Catholic family who had settled in Galway in the reign of Henry VI, although some accounts gave them deeper roots.

It was a difficult time for Ireland – and especially for Catholics. After the restoration of the monarchy following the death of Cromwell, Ireland became a land of great Protestant estates and small towns decaying under British trade restrictions. Brave priests had to celebrate their Masses among the ruins of churches and monasteries. At such a time, Kirwan was lucky to be born into relative privilege.

Luke Hutton, Kirwan lost his father at a young age. He was sent to live with his grandfather (who had once fought in the army of James II), and was tutored by a Dominican friar, beginning his fascination with theology. Catholics at this time were excluded from universities in Ireland – and, indeed, in Britain – so in 1750, Kirwan was sent to university in Poitiers, in France. He resisted learning French until one of his tutors discovered him reading chemistry books. The books were confiscated and replaced with texts in French, whereupon his grasp of the language improved dramatically.

By 1754, aged twenty-one, Kirwan had entered a Jesuit novitiate at Saint-Omer in France. The following year, however, his life took the first of many peculiar twists. His elder brother, the heir to the family estate, was killed in a duel. As the oldest surviving male, Kirwan was forced to give up his studies, and was called home to run the estate.

Back home, he determined to study law, but his Catholic origins were once again an obstacle for him. He had to foreswear his religion and adopt the faith of the Protestant Ascendancy. He accepted this, but the sense that he had been forced to adopt a foreign faith must have galled him.

In 1757, he married a local girl: Anne, daughter of Sir Thomas Blake of Menlo, in County Galway. But now he was struck by another bizarre misfortune. Kirwan did not know that Anne had run up debts that vastly exceeded her dowry. On their marriage, Kirwan became immediately responsible for his new wife's finances, and he was hauled away to gaol, where he had to remain until the debts were paid. Still, he stuck by Anne. They had two daughters together before she died in 1765, after just eight years of marriage.

Kirwan was eventually called to the Irish Bar in 1766, aged thirty-three. Owing to the ample means with which his family background provided him, however, he practised the law for only two years beore giving it up for his science. He built up a large library of chemistry books and, in a self-constructed laboratory, began to run careful and laborious experiments. He became fascinated by mineralogy. He carried out experiments on the properties of carbon in coal, and wrote essays on the analysis of soil, publishing the first systematic work on the subject in English.

By now Kirwan suffered from dysphagia, a difficulty in swallowing caused by problems with nervous or muscular control. He could never eat comfortably, as swallowing would induce convulsive movements and facial contortions. As a result he kept his diet simple – usually nothing but milk and ham – and he always dined alone, whether in his own home or even visiting friends.

In 1777, Kirwan travelled to London. Here he was honoured. In 1780 he was elected to the Royal Society, and he received the Copley Medal, the Society's most prestigious award, for his first scientific publication, the result of eighteen years' experimental work on chemical bonds. His best-known work, 'Essay on Phlogiston', was published in 1787. Unlike Hutton, Kirwan proved himself ultimately to have a flexible mind on the subject, publicly acknowledging the final 'subversion of the phlogiston hypothesis' by Lavoisier's debunking.

Given the impediments of his origin and social difficulties, Kirwan's success is a tribute to his ability and determination. Perhaps he was unhappy, however, that he had to go to England to cement his reputation. He must have felt once again like a subject of a conquered province, making his way to the capital of Empire to seek the approval of his rulers.

After ten years in London he determined to return to Ireland. Settling in Dublin, he strove to build the intellectual respectability of his homeland; just as Hutton helped to found the Royal Society of Edinburgh, so Kirwan helped establish the Royal Irish Academy, eventually becoming president. He was struck by yet another peculiar misfortune on his return. An American privateer, prowling around British home waters, intercepted a ship carrying Kirwan's library back from England. The library was taken back to Salem, where Kirwan's precious books were instrumental in the education of Nathaniel Bowditch, a self-taught American mathematician and astronomer.

In Ireland at this time, things were stirring. When the French Revolution erupted in 1789, a remarkable, if temporary alliance was forged between the Protestants and Catholics of the Irish intellectual elite. A series of radical political clubs were formed, called the Societies of United Irishmen. Perhaps it was Kirwan's own experience that drew him to support these rebellious figures. Like Hutton, Kirwan was a figure from Britain's subsumed Celtic fringe who was showing his patriotism in his own way.

By the 1790s, a widower for a quarter of a century, Kirwan had become a strange but weighty figure. He was settling into a life of strict routine. He liked to rise very early, and to retire early. If you wanted to visit him you were obliged to follow certain rules: on some evenings he allotted a particular time for visits, after which the door knocker was removed. He had become morbidly fearful of catching a cold. He would trot along the street, trying to minimise the time he spent out in the chill air; if you wanted to speak with him you had to run alongside him. Even indoors, he wore a huge cloak and hat and several woolly scarves – it was this habit that would eventually kill him. It is this peculiar, oddly wistful figure, bewigged, his neck swathed in a white scarf, that peers out of his portrait for the Royal Dublin Society.

Academically powerful, he was motivated by a deep and rare religious conviction, and a righteousness fuelled by the experiences of his life. In his geology, Kirwan had always been drawn to the Neptunist theories of Werner and his predecessors

– but born into the deep conviction of Catholicism and now educated in the intellectual tradition of Protestantism, he had become a Christian of a fundamentalist stripe, and he clung to the scriptural account of creation.

Hutton's ideas appalled him.

Kirwan's first anti-Hutton salvo was a paper called 'Examination of the Supposed Igneous Origin of Stony Substances'. Thirty pages long, he read it in February 1793 to the Royal Irish Academy in Dublin. The 'Examination' was a bold attempt to demolish Hutton's geological theory, which was quoted at length and attacked point by point.

Kirwan cited evidence from chemical experiments on rock samples to try to prove that granite and other rocks (except obviously volcanic ones) could not have been created by heat, an idea that was a 'peculiarly unhappy' aspect of Hutton's work. Kirwan then denied that all soils originated from erosion. He argued that not all the soil gets washed away to the sea, much of it being deposited along river banks or at their mouths. Whereas Hutton had claimed that much of the Earth's crust was composed of strata of sand, gravel and chalk, Kirwan said that the base rock of the world was granite, as that rock was the first to be laid down in the sequence of creation – just as Werner insisted. These assertions, untestable except by examination of the rocks themselves, must have wounded Hutton the field geologist.

Kirwan went on to attack Hutton on the mysterious nature of Earth's inner heat, which he called a gratuitous assumption. How could a fire burn within the Earth without fuel or air? He went through possible fuel sources, like sulphur, coal and bitumen – but even if any of these existed, you could hardly have fire without 'vital air'.

Kirwan's most devastating attack, however, was religious.

Kirwan was troubled by infinities, as philosophers always have been. Infinite regressions lead to paradoxes. Suppose you imagine the Earth is flat, and resting on the backs of four elephants. Fine, but what are the elephants standing on? Four more? OK, but then what? Is it pachyderms all the way down? Kirwan saw a similar infinite regression in Hutton's cycling worlds. Hutton (so Kirwan claimed) believed that such a succession of worlds must have existed from eternity, but a succession without a beginning was a logical paradox, and therefore to be dismissed. And besides, the notion of a cycle of infinite worlds clearly harked back to the old Greek ideas of an Eternal Return. The Greeks had been pagans: Hutton was therefore trying to revive anti-Christian ideas.

His notion of the succession of worlds, proclaimed Kirwan, was 'contrary to reason and the tenor of Mosaic history, thus leading to an abyss from which human reason recoils'.

This hit Hutton hard. As Playfair noted, 'the attack was ... rendered formidable, not by the strength of the arguments it employed, but by the name of the author, the heavy charges which it brought forward, and the gross misconceptions in which it abounded.'

Hutton knew he had to defend himself. Even more so than in 1785, Britain in 1793 was not a good place to be called a heretic.

In France, the Reign of Terror had touched everybody.

The great chemist Antoine-Laurent Lavoisier was just fifty. He was well known to the Edinburgh circle: Sir James Hall had met him in the 1780s and had been greatly inspired, but the Revolution had ended their contact. Lavoisier had been active in support of the Revolution, but he had fought hard to stop the suppression of the Academy of Sciences and other learned bodies – in the end he fell out, fatally, with Marat. At his trial, it was said, Lavoisier appealed for just a little more time to complete some scientific work. The judge replied, 'The Republic has no need of scientists.' The great mathematician Joseph-Louis Lagrange would say, 'It took them only an instant to cut off that head, and a hundred years may not produce another like it.'

Lavoisier's headless corpse was thrown into a common grave. Much must Hutton, Black, Hall and others have mourned Lavoisier's fate, and shuddered.

For some, it all seemed a great betrayal of the heady promise of the Revolution: 'Bliss was it in that dawn to be alive,' as Wordsworth would write, 'but to be young was very heaven.' Young James Watt, the twenty-year-old son of the engineer, had been enthused by France's great celebration of liberty, but became foolishly mixed up with French politics. He was censured for carrying the British flag into the revolutionary assembly. It has been said that Robespierre assailed him as a British spy, after which he was forced to leave Paris. The more mundane truth is probably that he had to leave on business. Watt senior was not impressed with this youthful folly, and must have been relieved when young James came home in 1794.

The Reign of Terror provoked a strong reaction in Britain. Scotland, nearly a century after the Union, was not notably democratic. Thanks to a tough property qualification, fewer than one man in twenty (and no women, of course) had the vote. The affluent new middle class in Edinburgh and the other cities, created by the economic growth since the Union, had no political voice. Power still rested with the lairds and landowners – and with Henry Dundas, the London government's representative on Scottish affairs. 'King Harry the Ninth', as he was known, ruled through patronage in the church, the law, academia and politics.

In the wake of hard times in Scotland – strikes in Glasgow against high food prices, a failed harvest in 1792 – there had been some agitation for enhanced democracy. In July 1792, some Edinburgh citizens formed a Scottish Association of the Friends of the People, advocating a Britain-wide programme of reform.

Now – with revolution and terror exploding in what had been the most aristocratic and absolutist of European nations – life was suddenly serious, and a sense developed that one's hold on position, power and wealth was only conditional. Among the circles of the Enlightenment, it no longer seemed so clever to flaunt wealth and excess, or to play at sedition, however subtly. The clubs, with their overtones of fraternity, suddenly seemed to have sinister undertones: even the Beggar's Benison, after all, was playfully Jacobite. Freemasonry came under suspicion, and for a time the opening of new lodges was forbidden.

Any signs of unrest among the masses had to be briskly stamped upon. In a letter to Black, James Watt senior wrote: 'The Rabble of this country are the mine of Gunpowder that will one day blow up and violent will be the explosion.' There was a massive crackdown on every suspected subversive group. William Pitt, the Prime Minister, suspended *habeas corpus*. Radical campaigners, such as proponents of extended suffrage, were regarded as seditious. In 1795 one 'traitor' was publicly beheaded in Edinburgh High Street: on that blood-splattered day, the glories of the Scottish Enlightenment must have seemed transient indeed.

The Church was central to Britain's social structure, and any challenge to it could be seen as a threat to the interest of the government and the propertied classes. Edward Gibbon was very aware of this. His *Decline and Fall of the Roman Empire* had made it difficult to admire many of the greatest saints of the early Christian Church. He was very unkind, for example, to St Simeon Stylites, whose claim to fame was spending thirty years on top of a twenty-metre column in the desert: Gibbon compared Simeon's absurd career unfavourably to those of the great 'pagan' thinkers of earlier times, like Cicero. Gibbon learned to be cautious, for there were statues of the realm available to punish anyone who attacked Christianity. In the midst of all this, Kirwan's charges of atheism and heresy against Hutton were disturbing indeed.

Ill as he was, the very day he read Kirwan's attack, Hutton turned to the only means he had of fighting back: his much-postponed book-length exposition of his geological theory. Hutton, the bedridden invalid, grew ever more gaunt, obsessively writing, writing, writing, as if he could prove with words what he could no longer demonstrate with his rocks.

18

'A passage from one condition of thought to another'

Hutton's book would be called *Theory of the Earth with Proofs and Illustrations*.

He planned it in four parts. By September 1795 Hutton had had printed all 1187 pages of his first two volumes, and lacked only the six illustration plates to complete them – one of these was John Clerk's charming image of the Jedburgh unconformity. His Edinburgh publisher would be William Creech, the old friend who had published Burns's poetry. The volumes would be sold at fourteen shillings in boards, and nine shillings and eight pence wholesale.

With the *Principles of Knowledge* and *Theory of the Earth* both still in preparation, Hutton was well enough to give a series of readings to the Royal Society of Edinburgh on his theories of physics. He gathered these into another book, *A Dissertation Upon the Philosophy of Light, Heat and Fire*, published in 1794 (in fact it appeared before the *Theory of the Earth*). Here he gave a full statement, at long last, of his theory of heat as an aspect of 'solar substance'. It is a cruel irony that in the same year that Hutton's phlogiston-tinged tome was published, the founder of modern theories of combustion was brutally executed by the Terror.

This work seems not to have attracted much attention, but it was another crucial element of Hutton's geological theory. With this book, his *Principles of Knowledge* and his writings on geology, Hutton was endeavouring to produce a complete and consistent body of physical theory and epistemological methodology to support his assertions about the Earth: his non-geological works were just as important to him as the overtly geological. It is a peculiar tragedy that his writings were so long and impenetrable that nobody at the time really understood what he was up to – and indeed only in recent decades have modern scholars managed to piece it all together.

Before the end of 1795, the first two volumes of the *Theory of the Earth* were published at last. The first included a reworking of Hutton's 1785 paper, with additions and expansions. Using material from other old essays, he presented his ideas on the origin of so-called 'primitive' rocks like granite and limestone. He cited specifics, like a bit of limestone he had found in a quarry in Wales which had borne the impression of a sea creature.

He went on to a new piece of evidence for his theory of underground heat: the existence of coal and bituminous strata. These strata, laden with hydrocarbons, are found all over the world, interleaved with other strata obviously formed at the bottom of the sea. But Hutton's opponents had argued that the bituminous material had infiltrated pre-existing strata, turning the rocks to coal. Hutton railed: 'The wonder now is how men of science, in the present enlightened age, should suffer such language of ignorance and credulity to pass uncensured.'

Above all, Hutton expanded on his theory of the cycling Earth, quoting his dramatic discoveries of the unconformities at Jedburgh and Siccar Point.

His second volume was a discussion of the evolution of landforms. Hutton gave a beautiful description of a fertile plain being destroyed by an encroaching river, and explored the example of the English Channel, bounded by chalk cliffs that are being steadily eaten away. He speculated that since England and France were obviously once joined it might be possible to work out from the present rate of erosion of the Channel's walls how long ago they had been split (the answer would have been around 80,000 years, which is of the right order of magnitude). This volume is now regarded as a foundation of the modern discipline of 'geomorphology', the science of Earth's surface features.

The *Theory* was the forum that Hutton used to fight back against Kirwan's attack. Hutton wasn't impressed with Kirwan's talk of experiments on bits of granite. He pointed out that he had been 'anxious to warn the reader' against imagining you could compare the results of subterranean heat, with its tremendous intensity, temperature and compression, with anything you could emulate on the surface of the Earth. 'Yet, notwithstanding all the precaution I had taken,' he griped, 'our author [Kirwan] has bestowed four quarto pages' on exactly such false comparisons.

Hutton also replied briskly to Kirwan regarding the source of his inner heat. He simply put the question aside, as he had in 1785. He had never claimed that his heat was based on fire. Besides, his heat didn't really *need* a source. Endlessly circulating within the body of the Earth, it was a dynamical fluid always available to repair the substance of the Earth. Kirwan would later call this dismissive response 'wholly paradoxical'. But Hutton knew he was on thin ice here, and in his long, rambling, repetitive chapters, he lost his way a little. In one place he even unwisely suggested that the heat might after all be fuelled by coal, produced from a previous geological cycle's plant life. His nerve had momentarily failed; better to have stuck to the line that the source of heat was simply outside his scope.

As for the paradox of the 'eternal' world, here Hutton was able to hit back hard. He had been misunderstood. He had never claimed that the world is eternal. He had only said that we have a limited ability to reconstruct Earth's past or foresee its future. Earth surely had a beginning and will surely have an end, but it was beyond his methodology to see them.

Hutton closed his first two volumes with a last statement on the divine design of the globe, in language that once again recalled Aristotle: 'We live in a world where order everywhere prevails and where *final* causes are as well known, at least, as those which are *efficient* ... Thus, the circulation of the blood is the *efficient* cause of life, but life is the *final* cause, not only for the circulation of the blood but for the revolution of the globe' (my emphasis). After all this time he had come back to the themes of the medical studies of his twenties – the *microcosm* and *macrocosm*, the metaphorical unity of body and world.

Hutton planned two more volumes. We know from surviving manuscripts (see the Epilogue) that there would have been at least ten chapters, six of them drawing on essays he had written between 1785 and 1787. These included write-ups of his visits to Glen Tilt, Galloway and Arran, essays on his reading of Alpine travelogues, and reports of the mineralogy of the Pyrenees, Calabria and Sicily, based on his readings. But there was a problem. Unlike the first two volumes of the *Theory*, the latter segments required many engravings, mostly based on drawings by John Clerk. As 1796 wore on, Hutton's health deteriorated and his medical bills mounted. He was running out of money, and the process of engraving was halted.

In the end, the third and fourth volumes were not published in Hutton's lifetime. The *Theory of the Earth* as it eventually reached the public was only half of what Hutton had intended. And it wasn't enough.

Hutton's two volumes actually attracted less attention than the 1788 paper had done. Hutton was known to be seriously ill, and wasn't raising any new issues: this was old news. And anyhow, the work was infuriatingly difficult to read. Lengthy as it is, the *Theory* is a hasty work – the old saw that it takes more time to write a short book than a long one was never better demonstrated.

Hutton has a reputation for being notoriously unreadable. I don't find his style in his formal papers all that bad, even if his prose sometimes reads as if it has been translated from another language, and Hutton the scientist certainly isn't as engaging as the scurrilous rogue we glimpse in his letters. It can't be denied that the most cogent account of his theory is the Abstract, the most unreadable of his multiple volumes: the shorter the better. But the modern legend of unreadability is surely compounded by a reluctance by some later mythologisers to recognise the logical basis of what Hutton was saying: that his theory was not a modern scientific argument proceeding primarily from the data, but an argument from final causes.

Hutton didn't even manage to convince his friends. To save a little money, he had sent them copies of the books as loose sheets, which they had to pay to have bound.

In December 1795 Watt sent him bits of granite from Cornwall, odd mixtures of metal ores, and stones from the Wrekin and the Malvern hills. Watt had been working on mechanical contrivances for 'pneumatic medicine', gadgets intended to cure illnesses – especially tuberculosis, from which his own daughter had died – with a mixture of gases. But there was nothing even the great engineer could build to help his old friend: 'I wish I could contrive remedies for your several complaints, but alas I cannot cure my own, though much lighter!'

Later that month, Watt wrote to Hutton from a tour of North Wales, bemoaning the lack of experimental confirmation in Hutton's work. 'Without pretending to be a believer, I see much to commend and admire. [But] I do not believe even in mechanics without experiment to which test I wish to bring all theoretical opinions if possible & so I should have yours served.' Watt continued with speculations about experimental geology, and about the rocks of Wales. 'I have some more facts I mean to trouble you with at intervals, if you think them sufficiently interesting. In the mean time I shall be glad to hear from you especially, how you continue to support your very cruel disease, most sincerely praying an alleviation of your sufferings & begging to be remembered to Dr Black & other friends. I remain, Dear Sir, your affectionate friend, James Watt.' It was the last letter Watt wrote to Hutton, and Hutton did not reply, though Black kept Watt informed of Hutton's condition.

Still Hutton's opponents assailed him. In October 1796, his old adversary Jean André Deluc published a forty-four page, three-part review of *Theory of the Earth*. It was a ferocious assault. Geology could never be disentangled from the sacred history, wrote Deluc. There was no evidence for a succession of worlds except in Hutton's imagination; Hutton's 'theory' wasn't an argument about nature but an invention *specifically intended* to attack Moses' holy account.

Now Richard Kirwan returned to the fray. He prepared a series of three essays 'On the Primitive State of the Globe and Its Subsequent Catastrophe' for his Royal Irish Academy, and read the first of them on 19 November 1796. Kirwan's new work amounted to a magnificent statement of a new Neptunist theory of the Earth.

It was obvious, he argued, that much of the globe was once in a soft or liquid state, dissolved in a uniform and chaotic ocean. The materials had settled out in orderly layers. The universal ocean had subsided and the continents emerged: it had been a once-and-for-all formation event. There was no room for Hutton's cycles, and certainly no evidence for them.

Kirwan tackled the old problem of the corpses of marine

animals being found buried in the rocks at the tops of mountains. Sea creatures hadn't existed when primitive mountains were formed – so the fossils must have been put there by a *later* inundation. That also explained the bones of elephants (in fact mammoths) and other tropical beasts that had been found in Siberia. Kirwan imagined Noah's great Flood starting in the southern hemisphere and rushing north: the animals must have been swept from Africa to Siberia by the deluge.

He regretted the way 'recent speculations' had shaded into atheism and infidelity. For Kirwan, the purpose of science was, as it had been for Burnet, as it always had been, to find an explanation of the world to match the ultimate truth, the revelation of scripture. If properly pursued, he said triumphantly, 'geology naturally ripens, or (to use a mineralogical expression) *graduates* into religion'. And Kirwan's own theory of the Earth was, unlike Hutton's, fully in accord with Old Testament writings: Kirwan wrote proudly, '[My] account of the primeval state of the globe and of the principal catastrophes it underwent, I am bold to say, [is as] Moses presents to us.' Kirwan's theory was a tour de force of theological science, perhaps the last of the magnificent edifices to be erected by the Biblical system-builders – and it was the utter antithesis of everything Hutton had argued for.

Hutton and Kirwan never met. But these two figures, both eccentric in their different ways, had become engaged in a war fought through journals and speeches and books. It was a feud to which the logic of their lives had led them.

Kirwan was a working scientist, and on one level his riposte to Hutton was an overdue scientific test of his ideas in the crucible of a peer review. But the argument between Hutton and Kirwan was at heart not just about science – and perhaps not really about science at all: it was about differing theories of the nature of God. *That* was why Hutton was so hurt by any implication or charge of atheism, aside from any trouble it could cause him or his friends. He was anything but atheistic; it was just that his view of God was radically different from Kirwan's.

And the feud was bitter. If Kirwan complained that Hutton's vision of 'no vestige of a beginning' led to 'an abyss from which human reason recoils', Hutton had responded by saying that 'the abyss from which the man of science should recoil is that of ignorance'.

Kirwan must have anticipated a fresh response from Hutton, and perhaps relished the prospect of another round of the great battle. But it was too late.

In 1796 Hutton suffered another relapse. 'He was again saved from the danger that immediately threatened him, but his constitution had materially suffered, and nothing could restore him to his former strength.' Confined to the house, emaciated and in much pain, Hutton became weaker. But his mind remained active. He continued to receive his friends, and he tinkered with work.

After his publication of 1795, Hutton devoted much of his remaining efforts to what would have become another immense book, his thousand-page *Elements of Agriculture*.

Though this sadly never reached a wide public, Hutton set out much wisdom in it: on crop rotation, the effect of climate on plant growth, soil fertility, farm management, fertilisers, animal husbandry and other topics. He also united his agricultural thinking with the broader themes of his philosophy. Through agriculture, human science cooperated with the mechanisms of nature to further God's purpose, that Earth should serve as a fertile, habitable globe.

Though Hutton had been reluctantly driven to farming, he was always proud of his achievements at Slighhouses. In the 1788 publication of his 'Theory of the Earth', he listed his honours as MD (Doctor of Medicine), FRS Edin. (Fellow of the Royal Society of Edinburgh) – and Member of the Royal Academy of Agriculture at Paris. It is a mark of Hutton's agricultural reputation that he was one of just twenty-three foreigners, drawn from across Europe and America, chosen as members of this prestigious and influential society. (The most famous member of all was probably 'M. le general Washington'.) Though in Britain he was known as 'the famous fossil philosopher', to quote Watt, in France Hutton had undoubtedly built up a reputation as an agricultural expert, and even received visitors on the subject. Unfortunately, all records of Hutton's involvement with the Academy were lost in a disastrous flood in 1910, which destroyed the Academy's archives.

In a sense, Hutton's thinking had once again come full circle. He knew that his book would not attract a large audience, if it was ever published at all, but he worked on it for his own satisfaction: writing was a comfort to him, as well as a means of expression. He took pleasure in returning to a subject which had been 'in a manner the study of [his] life'.

His health was worsening, and *Elements* bears the scars of exhaustion and illness. Like his other late works, it is strung together from earlier essays, leaving it rambling, repetitive and poorly structured. On some pages of the manuscript, you can see pencil marks where Hutton meant to reorder the paragraphs. But the lines are unsteady, a sign of his increasing weakness, and they often break off halfway through. It seems likely that he never even read the dictated manuscript all the way through.

The last chapter stumbles to a halt, mundanely, in the middle of a discussion about potatoes.

Hutton's friends tried to support him. They still wrote to him, sending him interesting samples and geological chit-chat.

Hutton himself continued to contribute to the fledgling Royal Society of Edinburgh, on the questions of written versus spoken languages, the geology of Arthur's Seat, and 'The Flexibility of the Brazilian Stone': 'No quality is more inconsistent with the character of a stone than flexibility. A flexible stone, therefore, presents an idea which naturally strikes us with surprise ...'

A paper on 'The Sulphurating of Metals' – to do with his phlogiston theories – was read for him on 9 May 1796. In their 'History of the Society' notes, the editors of the *Transactions* write, 'Such are the ideas which Dr Hutton had formed on the sulphuration of metal, and the theory by which it must be explained; and they are rendered more interesting, by being the last communication made by that ingenious and profound philosopher.'

In the spring of 1797, Hutton was pleased to receive the third and fourth volumes of De Saussure's travelogues of the Alps, to the earlier parts of which he had referred extensively in his *Theory*. It was like one last field trip, conducted in the mind.

Even in these last days, despite the critics, he may have continued to enjoy exploring his own ideas. Playfair writes that, 'With [Hutton's] relish for whatever is beautiful and sublime in science, we may easily conceive what pleasure he derived from his own geological speculations. The novelty and grandeur of the objects offered by them to the imagination, the simple and uniform order given to the whole natural history of the Earth, and, above all, the views opened of the wisdom that governs nature, are things to which hardly any man could be insensible; but to him they were matter, not of transient delight, but of solid and permanent happiness ... No author was ever more disposed to consider the enjoyment of them, as the full and adequate reward of his labours.'

On Saturday 26 March Hutton woke in a good deal of pain. He tried to work, making some notes about a new naming system for minerals. But the pain worsened and in the evening he started to shiver. His doctor, the younger James Russel, was called.

Hutton had written in the *Principles of Knowledge* that death was no longer to be seen as a disconnection between mind and reality, but simply a change in the way the mind perceives things: 'We are to consider death only as passage from one condition of thought to another.' Perhaps that comforted him. But Hutton, the man who had given Earth a history deep beyond human imagining, had now himself run out of time.

As Playfair reports, Russel arrived too late. Hutton used his last strength to reach out his hand towards his doctor.

THREE

Uplift

'I could see the marks
of his hammer'

Hutton's friends were devastated. Playfair wrote of 'the consideration of how much of his knowledge had perished with himself, and ... how much of the light collected by a long life of experience and observation, was now completely extinguished'.

In October 1797 Playfair visited Arran. He clambered gamely around the island – 'I can endure a little hardship when put in the balance against knowledge' – and studied basalt dykes and contacts between granite and schists. Playfair wrote to Sir John Hall, 'The junctions I saw were all visited by Dr H. At one of them I could see the marks of his hammer, (at least I thought so), and could not without emotion think of the enthusiasm with which he must have viewed it. I was never more sensible of the truth of what I remember you said one day when we were looking at the Dykes in the Water of Leith, since the D's death, "that these Phenomena had now lost half their value".'

There would be no neat resolution to Hutton's arguments with Kirwan and his other critics. But if the man had died, his ideas lived on – and the battle over them was only just beginning.

In the year of Hutton's death, for example, the third edition of the *Encyclopaedia Britannica* weighed in with a twelve-page critique. The *Encyclopaedia* attacked what it saw as Hutton's claim that the world was eternal, and stated that the world was full of evidence for the Flood. Hutton had even got his chemistry wrong, said the *Encyclopaedia*. There were other reactions from the British Isles, and several from the Continent – including one from Hutton's old foe Faujas, who dismissed Hutton's thesis as 'a memoir containing general views of the subject [rather] than a body of observations'.

In the face of these attacks, Playfair, Hall and others decided to carry Hutton's arguments forward for him: they would amass evidence where he could not; they would speak, where he was silent.

Sir James Hall had never quite bought Hutton's theory: 'I must own that on reading Dr Hutton's first geological publication I was induced to reject the system entirely,' he said. There had followed 'three years of almost daily warfare with Dr Hutton' over the theory, which at last Hall had begun to view 'with less and less repugnance'. Hall remained fascinated by the idea that you could explore geological ideas in the laboratory, and Hutton's death gave him a poignant opportunity to continue this work.

By 1790, a heat origin for granite had become crucial to Hutton's argument. But Hall as a chemist knew that there were problems with the idea that granite had solidified out of a melt. Actual experiments on melting granite and cooling it again had resulted not in the crystal-laden stuff you found in the field but in a shapeless glass. Similarly basalt, when melted and cooled in the laboratory, didn't look much like field samples either.

In 1790, Hall had visited a glass factory in Leith. There was an accident, and Hall happened to see what happens when glass cools slowly. Under these conditions glass can *crystallise*. Perhaps, Hall thought, if you cooled molten granite slowly enough you would recover not a glassy mess, but a crystalline structure. It was an obvious thing to test in the lab. That same year, Hall gave a paper to the Royal Society of Edinburgh in which he discussed Hutton's field observations on granite and basalt, and referred to well-known experiments on melting crushed feldspar and quartz. He went on to speculate about how crystals of different types could be produced by differing cooling rates in the bowels of the Earth. To Hall, this survey was just the start of what could have become a programme of laboratory tests of Hutton's ideas.

But Hall had encountered unexpected hostility from Hutton himself.

The trouble was, Hutton had come to believe that the conditions of great temperature and pressure that surely prevailed deep in the Earth were out of reach of any conceivable apparatus. He censured those who 'judge of the great operations of the mineral kingdom, from having kindled a fire, and looked into the bottom of a little crucible'. He may have been nervous that negative results from flawed experiments would do his arguments more harm than good. Hall's reply was that 'the imitation of the natural process is an object which may be pursued with rational expectation of success'.

Nonetheless, out of respect for the older man – in 1790 Hutton was sixty-four, Hall a mere twenty-nine – Hall set aside his experiments. 'I considered myself bound, in practice, to pay deference to his opinion, in a field which he had so nobly occupied, and abstained, during the remainder of his life, from the prosecution of [my] experiments ... [But] in 1798 [after Hutton's death], I resumed the subject with some eagerness, being still of opinion, that the chemical law which forms the basis of the Huttonian theory, ought, in the first place, to be investigated experimentally.'

Hall began by melting bits of Edinburgh basalt in the furnace of an iron foundry. If the basalt was allowed to cool rapidly it formed a black glass, but, in a dogged series of tests, Hall managed to slow the cooling, and by his twenty-seventh try he had indeed managed to produce a crystallisation that looked very like natural basalt. While the arguments continued over Hutton's theories, in the laboratories Hall continued his patient exploration.

Playfair was always the leader and the most ardent of the new 'Huttonians'.

His first task, Playfair decided, was to produce a biographical memoir of Hutton, incorporating reflections on his work. He began to gather material from Hutton's friends and family. He consulted Black, though the ageing chemist had by now retired from his teaching post. Hall provided letters Hutton had written to Hall's father. Isabella described her brother's early life. The Clerks surely reminisced to order.

In the process of researching his biography, though, Playfair reread Hutton's book-length *Theory of the Earth*, and was appalled at how badly written it was. In an age in which the elegance of a theory's expression added greatly to its chances of a good reception, Hutton's foggy impenetrability was going to be a serious obstacle to his acceptance. So Playfair – urged on by Robertson, who had himself helped during Hutton's literary struggles – took it on himself to rewrite the theory.

Playfair presented the result to the Royal Society of Edinburgh in June 1799, and published it as *Illustrations of the Huttonian Theory of the Earth* in 1802. Playfair's logical restructuring of Hutton's material certainly makes it a lot more readable and cogent. To counter Kirwan's attacks, Playfair added a series of 'notes', in fact amounting to hundreds of pages. He quoted new examples of unconformities in Scotland, Yorkshire and Cumberland, and more granite intrusions.

Playfair dismissed the decay-ridden theories of Burnet, Buffon and others as irrational: 'The death of nature herself is the distant but gloomy object that terminates our view, and reminds us of the wild fictions of the Scandinavian mythology, according to which, annihilation is at last to extend its empire even to the gods.' By contrast, Hutton's theory was beautiful and inevitable: Earth was 'a system of wise and provident economy' designed for the support of life. It was surely the elegance and generosity of the theory, as well as loyalty to the man himself, that attracted Hutton such intense support from intellectuals like Playfair and Hall.

One of Playfair's audience that June day was Francis Horner, a friend of a friend. As Napoleon's war raged across Europe, and civilisation itself seemed under threat, Homer mused on the wider implications of Hutton's theory: 'And shall we have future theories of the moral history of the Earth ... tracing back many former transitions from civilisation to barbarism, and presenting, in the future prospect, an endless, irksome succession of the same changes?' (In fact, cyclical theories of history would indeed become fashionable in the nineteenth and twentieth centuries.)

The theory Playfair presented so skilfully, however, wasn't quite Hutton's.

Aiming for a wider acceptance, Playfair had subtly modernised his friend's work. He downplayed Hutton's rather old-fashioned reliance on arguments from design. He mentioned final causes and God's purpose a few times, but with nothing like the centrality and emphasis of Hutton's original. On the other hand, he emphasised Hutton's reliance on evidence. As he would write in his biography, 'from his first outset in science, [Hutton] had pursued the track of experiment and observation, and it was not till after being long exercised in this school, that he entered on the field of general and abstract speculations.' In fact, as we've seen, though Hutton's field-work was meticulous, his 'abstract speculations' had from the beginning done a great deal to guide his thinking.

Playfair also skated over Hutton's denial of historical change. Hutton had argued that the rocks showed the traces of previous cycles, but that these cycles were like ours in every important way. By now, however, the stratigraphers were coming up with what looked like clinching proof, through the fossils, that there *was* a trace of some kind of history in the rocks. So Playfair skipped over Hutton's ahistorical perfect world, and claimed that you *could* read a unique history in the rocks' 'marks of the lapse of time'.

In so intelligently recasting Hutton's eighteenth-century thinking in a nineteenth-century light, Playfair saved Hutton's ideas, and presented them to a new generation. Charles Lyell, the most influential geologist of the nineteenth century, would get his Hutton from the *Illustrations*. But the 'Hutton' of Playfair's retelling wasn't quite the original. From this point on, James Hutton himself began to slip deeper into shadow.

Joseph Black was also in Playfair's audience that day – but Black's time, too, was almost done.

In later years Black had suffered from frail health, and every time he caught a cold he found himself spitting up blood. He preserved his energies with a modest and abstemious diet. It was over such a meal that he would eventually die, reported his friend Ferguson: 'Being at table, with his usual fare, some bread, a few prunes, and a measured quantity of milk diluted with water; and having the cup in his hand when the last stroke of his pulse was to be given, he appeared to have set it down on his knees, which were joined together, and in this action expired, without spilling a drop – as if an experiment had been purposely made, to evince the facility with which he departed.'

If Hutton's friends were writing books about him, then so were his enemies.

In 1799, Richard Kirwan published his *Geological Essays*. This was nothing less than a book-length attack on Hutton, now two years dead. The formidable Irishman, reclusive, eccentric and obsessive as ever, ripped into Hutton's theory once again, as well as his attempts to answer Kirwan's own earlier objections.

The questions were familiar, but restated with authority. No experimenter had yet shown how chalky matter could be fused and crystallised. Kirwan didn't deny Hall's experimental results, but just because Hall had managed to persuade molten basalt to cool into a stony form, that didn't prove this was what went on in the ground.

Besides, Hall's tests were a distraction from the main issue, for they considered the *action* of the heat, not its *source*, which continued to be the subject of great controversy. Where was all this heat supposed to come from? Brushing aside Hutton's protests that his 'heat' was not necessarily the product of burning, Kirwan asked how there could be enough combustible matter or 'vital air' – oxygen – inside the Earth, to fuel such a mighty source of energy.

An even more serious attack about the question of heat was made by John Murray, who wasn't at all impressed by the idea of the Earth having a fixed reservoir of heat that circulated endlessly. Heat just wasn't like that, Murray argued: the planet would tend to a uniform temperature as the heat spread itself out, and then all the heat would dissipate into the air. Playfair responded by considering the analogy of a metal bar with one end held in a fire. So long as you maintained the fire, heat would flow along the bar, and it would remain hotter at one end than the other, never reaching a uniform temperature. Perhaps the Earth was like that, provided some central source of heat continued to heat up the inner layers.

Ultimately all this went nowhere. Hutton's model of heat – of circulating immaterial fluid – had nothing to do with the standard theories then being discussed. There could be no meeting of minds, and the heat debate would melt away in the face of the modern theories of thermodynamics developed during the nineteenth century.

Still the battles went on, as 'the Vulcanist and Neptunist war raged in the hall of [the Royal Society of Edinburgh] and in the class-room of the University'. These debates helped to generate interest in geology throughout the United Kingdom, and to establish the new Royal Society of Edinburgh's own reputation.

Hutton's friends were concerned, however. Hutton continued to come under attack for his supposed atheism, and the chances of shifting the opinions of the Neptunists, with their lingering theological fringe, were slender. It seemed possible that Hutton and his ideas would soon sink out of sight, like those of so many system-builders before him. It was as if a dead man was becoming more dead with time.

In 1803, Playfair presented his *Biographical Account of the Late James Hutton* to the Edinburgh Royal Society. The book version published two years later would become the primary source for biographical information on Hutton.

Even before this willing audience, Playfair faced a challenge. Hutton's primary achievement – his geological theory – was at a low point in its intellectual fortunes, and even his closest friends were prepared to admit that Hutton had been something of an oddity in Edinburgh's fashionable circles. His work seemed to hark back to a vanishing era of system-building in the absence of facts, a mode of discourse which was becoming increasingly discredited.

But Playfair did a magnificent job. He sketched what he could reconstruct of Hutton's early life: he didn't know much about Hutton's time as a farmer, and little about his early personal life, notably the scandal that had driven Hutton to the farm in the first place. He summarised the development of Hutton's theory, noting how important Black's chemical experiments had proved to be. And in a final, long section, Playfair gave an affectionate pen portrait of his friend as scientist and man: 'Dr Hutton possessed, in an eminent degree, the talents, the acquirements, and the temper, which entitle a man to the name of a philosopher ... He was upright, candid, and sincere; strongly attached to his friends; ready to sacrifice anything to assist them; humane and charitable.' Those who attended must mostly have known Hutton and been attached to his memory, and Playfair duly gave them a eulogy – even if his sanitised portrait didn't quite fit the facts.

It was effective, though. In the face of the continued hostility of the critic, it inspired a group of Hutton's friends to formalise their Huttonianism still further, and support his ideas more systematically. In death as in life, Hutton's story is one of friendship, loyalty and mutual support, among his friends at least – this had never extended, of course, to his son.

In June 1803, Playfair, Hall, Walter Scott, Francis Horner and others did what Edinburgh gentlemen did best: they formed a new social club. It was called the Friday Club, and would meet weekly at a tavern in Shakespeare Street; this genteel gathering would form the focus of Huttonian arguments for the next fifteen years. The Huttonians argued in print and in public with Hutton's critics, and they determined to go back into the field and into the laboratory to gather more evidence to shore up the theory. In 1807, for example, Playfair went back to Glen Tilt, still the cause of great controversy over granite, to conduct a more careful survey.

The Huttonians even used Edinburgh itself to try to win over more converts. They developed a habit of taking interested parties to the geological highlights of Edinburgh, as well as to sites further afield, like Siccar Point. One visitor remarked that Edinburgh, with the grandeur of its setting, would – unlike London, for instance – inspire even the most ignorant to ask questions about the rocks.

Sometimes these tactics were effective. A young geologist called George Silliman, torn between the Huttonian and Wernerian camps, would be swayed by the evidence: 'Though I am by no means a convert to this theory, I cannot but be impressed with sentiments of deep respect towards the author of it, a man of original genius who saw with his own eyes, who saw clearly, and knew what was worthy of being seen.'

Meanwhile, Hall continued his patient tests on heated and crushed rocks.

In December 1795, when James Watt wrote to Hutton from Birmingham to give his first reaction to the *Theory of the Earth* – 'I do not believe even in Mechanics without experimentation' – he had suggested a new way to test Hutton's theory. Watt knew a great deal about furnaces, and thought it ought to be possible to use a furnace to reach very high temperatures and pressures. He suggested sealing a sample inside a gun barrel along with a little water, and heating it in a furnace. The water's high pressure would simulate conditions inside the Earth. You could use such an apparatus to try to turn wood into coal, and chalk into marble, as Hutton predicted should happen in such conditions.

Hall eventually ran more than five hundred experiments along these lines. He would put his samples of limestone in paper cartridges inside common gun barrels, ram them full of clay, and then seal both ends with iron. Watt, it turned out, had underestimated the difficulties. Some of Hall's gun barrels failed – one of them wrecked the furnace he put it in – but he did succeed in getting a suggestive series of melts. He went on to try porcelain tubes rather than gun barrels.

Hall investigated what happened to animal and vegetable remains under such conditions, and became convinced that coal really was made up from such remains. The material was so reduced by the heat that he concluded the volume of coal to be found in the ground today must be just a fraction of that of the living matter once deposited on the surface.

Hall's most striking results came from tests on calcites – chalk, limestone, spar, marble, and the shells of sea creatures. He found that such samples could indeed be converted into a crystalline marble at high temperatures, and there was no doubt that they did indeed melt. In one test he would stand periwinkle shells upright inside his gun barrels; when he retrieved the samples he would find the remains of the shells, with rounded edges sticking out of a hardened mass, or else they would disappear altogether into the melt.

All this persuaded Hall at last that Hutton's ideas about the effects of heat must be essentially right, and that the Earth's inner heat – whatever its source and nature – must therefore exist. The Huttonians must have regretted that Hutton himself had been so resistant to such tests in his lifetime.

Hall was a true pioneer in the discipline that has become known as 'experimental petrology', in which rocks of different compositions are subjected to differing regimes of temperature and pressure, in an attempt to model what happens to them in the interior of the Earth. It only became possible to simulate extremely severe conditions during the twentieth century: by the 1950s, though, experimenters could model conditions that prevailed throughout Earth's crust, by the 1980s they had 'reached' depths of seven hundred kilometres – and more recently still, a miniature apparatus using diamond anvils has actually reproduced the conditions to be encountered *two thousand* kilometres deep, at the boundary between Earth's mantle and its core.

In 1804, James Watt's twenty-seven-year-old son Gregory, perhaps inspired by his father, decided to repeat some of Hall's experiments, but on a monumental scale. He took two hundred kilograms of basalt to an ironworks and tried melting it. What he got was a huge glassy mess, but it was a start.

Gregory was a talented young man. On meeting fourteen-year-old Gregory, Hutton had been fond of him: 'I was never more pleased with a young man,' he told Watt. 'He is everything that one could wish; he is both the gentleman & the philosopher. I make no doubt but he will succeed very well in business; & I hope his chymical knowledge will be of use to him.' But this was not to be. The same year he published his first results on basalt, Gregory died of consumption – much to the distress of his father, who had already lost a daughter.

Meanwhile, another of Hutton's old friends enjoyed a great triumph.

In May 2002, a historian at the National Maritime Museum in Greenwich was going through the papers of Admiral Lord Nelson. He chanced to turn over a long-disregarded list of names, written out in Nelson's own hand. On the back of this scrap is a scrawled map – really not much more than a doodle of dotted lines and sweeping curves. It doesn't look like much, but the historian immediately recognised its significance.

In 1805, at a dinner party on board the *Victory*, Nelson told his admirals of his battle plan for Trafalgar. He would describe their reaction as like 'an electric shock'. Some would even weep. And during the party he sketched his tactics on this bit of paper. A thick diagonal line represents the enemy fleet. But the British fleet is formed into three divisions and it cuts the enemy line in two places. In this hasty sketch you can sense Nelson's animation, and the tension of that dinner party; where the British line of attack cuts into the French fleet, Nelson's pen has dug deeply into the paper.

The plan worked, of course – and these were precisely the tactics John Clerk of Eldin had been working out with the cork models that had so fascinated Walter Scott. After the battle, the captain of the *Defiance* sent Clerk Nelson's memorandum to his commanders on how the battle was to be fought: 'Mr Clerk will perceive with great pleasure that the present form of battle is completely accordant with his own notions.'

As the years wore away, the Huttonians' persistence slowly paid off in establishing a community of support.

The new science of geology polarised into two camps, the Wernerians and Huttonians. The Royal Society of Edinburgh, and the newly founded Geological Society of London, leaned towards Hutton, but Werner had powerful disciples, even in Edinburgh. Both sides could point to supporting data, and both theories were consistent and well-founded. Not only that, the fringe of the Wernerian camp still contained many weighty figures, not least Kirwan, who clung to the Biblical chronology. To them, Hutton remained atheistic.

The stakes couldn't have been higher. The whole of the future of the science was going to be shaped by the outcome of the debate, and a lot of careers and reputations were going to be made or destroyed.

In the heat of the controversy, geology even became popular with the public. As the Romantic era dawned, writers like Wordsworth and Coleridge wrote magnificent hymns to the beauty of the landscape. Coleridge especially was something of a geology groupie. He corresponded with Erasmus Darwin on Hutton's *Theory of the Earth*, and the British Library still holds a copy of Hutton's *Principles of Knowledge* annotated by Coleridge himself.

Meanwhile, a new generation was giving geology a more practical bent. Roads and railways were crossing the continents, and in the great new industrial cities factories sprouted like mushrooms. All of this took resources. Giant quarries were opened to yield building stone and lime for cement, and metal ores and sources of fossil fuel were in great demand.

For the time being, the origin of the Earth was a lot less important than a decent map of what lay underfoot. This was the first great project of the Geological Society, founded in

London in 1807 with a self-proclaimed mission to focus on fact-gathering.

How could such a map be made? Britain's strata are notoriously folded and fragmented – 'topsy-turvy, twisted, crisped and curled', as Byron said. The answer occurred to the English canal surveyor William Smith.

Smith, born in 1769, was the orphaned son of a blacksmith from a village in Oxfordshire. In 1792 he had descended into mines in the county of Somerset. He was struck by the orderly layering of the strata through which he passed – and he was surprised by an unconformity every bit as startling as Siccar Point, though buried under the ground. Smith wasn't concerned about *how* the strata had got there. What mattered to him was that the basic principle of geological superposition – that younger rocks overlay older ones – could be the basis for some kind of map of the geological treasures to be found in the Earth.

To make such a map, you would have to be sure of the relative ages of your strata, which isn't always easy: minerals are minerals, whenever they are formed. If you want to date a rock, there is one rigid requirement: you need something about the rock that changes in a recognisable and irreversible way, so that each sample is left with a unique time-stamp. And Smith found his time-stamp in fossils. The different strata, no matter how similar in their basic rock types, always contained different fossil fauna, and that was how you could tell them apart.

Smith's final publication was in 1815, with a tremendous and magnificent map of England and Wales, almost three metres high by two wide. A copy of Smith's great work still hangs today in Burlington House, home of London's Geological Society. The familiar outline of the country is filled in with swirls and dapples of colour, blue, grey, yellow, orange, red, umber. A great orange sweep from the Severn estuary across the country towards the Humber shows a belt of Jurassic limestone, and a vast semi-circle of grey in the south-east is the Cretaceous chalk. It is as if the country has been flayed of its people, cities, roads and greenery, to expose the underlying rocky bones.

Smith, the Oxfordshire yeoman, suffered from many financial

difficulties, and from snobbery. He suspected that the leisured gentlemen of the nascent Geological Society had plagiarised his map to produce their own inferior version (and so they had, shamefully undercutting Smith's price and driving him to debtors' prison). Later in life, Smith's achievement was amply recognised and rewarded by the British geological establishment, and his maps were the prototype of modern geological maps.

Hutton himself had realised the potential usefulness of a map of geological outcrops. In about 1770, he offered to make one of Scotland for his fellow geologist John Strange, and he had asked Watt to make one of Cornwall for him. (Hutton was not the first to conceive the idea – one Martin Lister had floated the notion as early as 1683.) As noted earlier, Hutton, conditioned by Scotland's ruined strata, never thought of using the sequence of strata to map geological time, and had never spotted the history implicit in the fossil-laden rocks.

As such maps were compiled, it quickly became obvious that the simple Wernerian division of the strata just wasn't adequate to describe the reality of the rocks. The Huttonians smelt blood.

From 1811, Playfair began to publish commentaries in magazines like *The Edinburgh Review*, leaning heavily on the results of the new stratigraphy, in a concerted effort to bring Hutton to the attention of a wider public.

Meanwhile, Sir James Hall gave the Royal Society of Edinburgh a spectacular demonstration of how strata could be folded up, just as Hutton predicted. He piled up layers of cloth on a table and weighted them down with an old wooden door and heavy weights. The cloth represented his strata, the weights the layers of rock over them. Hall set two vertical boards at the ends of the pile, and moved the boards towards each other by hammering them with a mallet – thus representing the great mechanical forces working within the Earth. The weighted door was raised up, and the cloth developed folds which resembled distorted strata of greywacke to be seen in Berwickshire and elsewhere. Hall, ever the practical man, even went on to build an automatic strata-folding machine.

By 1815, as Napoleon began his final exile, there had been a string of published reports more or less favourable to Hutton, and no replies of substance. Explorations of the volcanic island of Iceland showed the workings of Hutton's inner heat. Evidence for the igneous nature of basalt, granite and the rest continued to accumulate – and so did more direct evidence of the inner heat: from the rising temperatures observed if you went down a mine, for example.

The Wernerians at last began to lose ground. There was no decisive moment, no *Eureka*. As late as 1813, the Reverend Joseph Townsend launched a fierce attack on Hutton's religious position, in a book uncompromisingly entitled *The Character of Moses Established for Veracity as an Historian, Recording Events from the Creation to the Deluge*. Who was the better historian – Moses, who had spoken to God, or James Hutton, who only had rocks to consult? But the case was being made.

By now, though, Hutton's generation were fading into the shadows.

Playfair decided to produce a second edition of his *Illustrations*, in order to reinforce Hutton with new observations. While the war with France continued, Playfair travelled around Britain, and visited Ireland to see the Giant's Causeway. In 1816, once the turmoil following Napoleon's escape from Elba had passed, Playfair set off on a new round of journeying in Europe. He made first for Paris, where he met Cuvier and others. Then he visited the Swiss Alps to inspect the strange erratic boulders there. He correctly concluded that only glaciers could have transported such monsters. He stayed the winter in Rome, then moved on to Naples, where he saw Vesuvius in eruption. He crossed the Alps again, undertook two weeks of mountaineering near Lucerne, and visited the Auvergne district, with its extinct craters, basalt flows and domed hills.

It was already nearly twenty years since Hutton had died. Playfair was sixty-seven: it was too much, and he was exhausted. By the end he was making only sparse notes. He returned home. He never gathered the energy to produce his second edition of the *Illustrations*. He died less than two years later, in 1819.

Though as ever only a diffident advocate of Hutton's theories, Sir James Hall continued his careful experimentation, and his subtle support of his friend's legacy. In 1824, aged sixty-three, Hall accompanied yet another keen young geologist on a repeat of Hutton's classic expedition to Siccar Point. The youngster was much impressed – and, a quarter of a century after Hutton's death, it was to him that the responsibility for the next stage of the argument would devolve.

His name was Charles Lyell.

20

'It altered the tone of one's mind'

What was most important about Charles Lyell was that, unlike most of Hutton's contemporaries, he got Hutton's point about uniformitarianism. But Lyell applied Hutton's great idea with a ruthless severity that even Hutton would not have recognised.

In the past, said Lyell, Earth had always looked much as it looks now. The climate changed, of course, but that was caused by a shifting arrangement of land and sea, and followed great cycles. Everything concerning plant and animal life was determined: similar conditions recurring in the future would give rise to *exactly* the same species as in the past. For now, the northern hemisphere was in 'the winter of the "Great Year"', but one day the long summer would return, restoring vanished exotica with it: 'Then might those genera of animals return, of which the memorials are preserved in the ancient rocks of our continents. The huge iguanodon might reappear in the woods, and the ichthyosaur in the sea, while the pterodactyl might flit again through the umbrageous groves of tree-ferns.'

It was an astonishing, beautiful idea – and it was ferociously argued. To Lyell this was more than just science: he believed he was immersed in nothing less than an intellectual war.

After the death of Werner, Georges Cuvier, born a quarter of a century after Hutton, had become the most famous geological figure in Europe. Even during Napoleon's wars, Cuvier explored the rocks of the Paris basin, mapping the strata of that great bed of chalk. His central discovery (independently made by William Smith in England) was that different strata contained different assemblages of animals, which could therefore be used to date those strata.

Cuvier's thinking was deeper than that of Smith the map-maker. He tried to reconstruct what he could of the vanished animals by comparing their fragmentary remains to the bones of living creatures. He was able to demonstrate that many animals to be found in the strata – like mammoths, Irish elk and woolly rhinos – had no modern counterparts. With such careful analyses, Cuvier proved the reality of extinction.

However, the assemblages of life forms in the different strata did not give way to each other in an orderly way. Rather, there seemed to be sharp cut-offs. The changes were so precisely defined that Cuvier decided they must have been caused by violent and sudden catastrophes. He speculated that what had drowned the great Parisian menagerie might have been Noah's Flood itself.

Today we recognise several 'mass extinctions' in the fossil record – perhaps as many as eighteen. We're pretty sure that the great killing sixty-five million years ago followed the impact of a comet, a global spasm that ended the reign of the dinosaurs. The greatest death of all was a cataclysm 250 million years deep, when nearly everything was lost. We still don't know what caused it.

Cuvier's ideas resonated in the public mind. Europe was emerging from the Napoleonic era, a time of abrupt changes and devastating destruction. Perhaps the public were more ready for a geological tale that was a better metaphor for their own experiences than the slow creep of Hutton's uniform processes.

However, Cuvier's 'confirmation' of Noah's Flood, coming from such an esteemed authority, was meat and drink to that dwindling but vocal band of geologists who still hoped that geology could verify the Bible. Even that fickle Huttonian Sir James Hall speculated that a sudden uplift at the bottom of the sea might have caused the Deluge. In his characteristically practical style, Hall started setting off gunpowder explosions underwater to see if he could make a scale-model tsunami.

Charles Lyell was appalled by this revival of scriptural geology.

Softly spoken and reserved in manner, Lyell was born in Scotland a few months after Hutton's death in 1797. At the insistence of his father Lyell studied law. But during a summer vacation from his studies at Oxford, young Lyell read a copy of Robert Bakewell's *Introduction to Geology* – stridently anti-Wernerian and emphasising the immensity of time. Lyell had been a naturalist – butterflies and beetles – but soon geology was competing with his law studies.

Even by his mid-twenties, Lyell's own ideas had been starting to emerge. He had an aesthetic preference, he told friends, for 'the agency of known causes'. There was Deist thinking in his own background: the nature of God was to be inferred from the study of nature's order. Then, in 1828, Lyell visited Italy, Sicily and France. At the bay of Naples, he was very struck by the Temple of Jupiter Serapis, of which three columns remained upright. Each of the columns had been eaten into at a high level by marine bivalves. So the columns must have been lowered into the sea – and then raised again, through several metres, in historic times. Not only that, these great movements had been so gentle that some of the temple's thin columns had survived without being toppled over: gradual, uniform processes, working evenly through time.

Lyell became convinced that Hutton's explanations were on the right lines. The latter-day catastrophists had to be defeated if geology was to emerge, after all, as an orderly science that would not have shamed Newton. And that was what Lyell determined to do.

He published the first volume of the book that would become his masterpiece in 1830. From the very start, it was one long argument to support the cause of uniformitarianism (though that word would be coined by a rival two years later). Even the title could not have been more explicit – *Principles of Geology: Being an Attempt to Explain the Former Changes of the Earth's Surface by Causes Now in Operation*. To acknowledge Hutton's priority Lyell opened his work with a quotation from Playfair, but he admitted that he knew little of Hutton's work directly: 'Though I tried, I doubt whether I ever read more than half [Hutton's] writings, and skimmed the rest.'

Lyell's uniformitarianism was taken to a fanatical degree. He insisted that *every* past event had to be explained by causes now operating: there was *nothing* in the past that we cannot see in the world around us now. Causes, wrote Lyell, had *always* worked at about the same rate as they do now. Any disasters, like floods or volcanoes, only had local effects: there had been no global mishaps. He believed Cuvier's sudden transitions in the fossil beds were just artefacts of a poor record. The rocky narrative was like a book of which 'only here and there a short chapter has been preserved; and of each page, only here and there a few lines'.

Perhaps Lyell's powerful arguments were necessary to rid geology of floods and other Biblical hangovers, but his propagandising would lead to an unfair caricaturing of his opponents. Cuvier, for example, was no prisoner of theology; it was just that when he looked at the rocks, unlike Lyell, he saw an unmistakable record of cataclysm.

Lyell's work was a great success, and hugely influential. It was much admired for its discipline and sense of rigour, and did a great deal to establish geology on orderly foundations, and to promote Hutton's essential arguments.

In the end, though, Lyell's uniformitarianism would prove too strict a model for the messy realities of the rocks. Evidence continued to turn up, like the great U-shaped valleys of north Britain, that surely could *not* be explained away by Lyell's slow-acting processes, no matter how long they were given to work.

The tension continued until the late 1830s, when Louis Agassiz presented a new theory: not water, but ice. The geologists quickly realised that many anomalous features of northern Europe, like the flat-bottomed valleys and the erratic boulders, had been shaped by the action of ice, not water. The idea of an Ice Age was born.

Agassiz was actually a fan of Lyell, and praised his *Principles* as the most important work on geology to date. Lyell, on the other hand, had a great deal of trouble with Agassiz's theories. Agassiz was a catastrophist, and therefore an obstacle to Lyell's great project. It took Lyell until 1858 before he finally accepted Agassiz's glaciations.

In the end, though, it would not be rocks or ice but living creatures that would present Lyell with the greatest challenge of his intellectual life.

In 1831 – a year after the publication of Lyell's *Principles* – Charles Darwin, aged twenty-two, was preparing for an extraordinary voyage on a survey ship called the HMS *Beagle*. And the young Darwin took with him on the *Beagle* a copy of Lyell's newly minted *Principles*.

Darwin would later admit that before opening *Principles* he knew nothing of geology: 'As far as I know everyone has yet thought that the six thousand odd years [of Ussher] has been the right period,' he would write to his sister. When he read Lyell, he was drawn to the central truth of the Huttonian view. The clincher was his experience of an earthquake in Chile, when he actually witnessed uplift: on one island the *Beagle* crew saw rotting mussels stranded on rocks three metres above the high-water mark.

Darwin was able to see a major gap in Lyell's argument, however. The creation of species was the point where Lyell's application of uniform natural laws broke down. Darwin said, 'We can allow satellites, planets, suns, universe, nay whole systems of universes to be governed by laws, but the smallest insect, we want to be created by special act.'

Darwin marvelled at what he found in the tropics – 'The land is one great wild, untidy luxuriant hothouse, made by nature for herself' – and he came home with the raw material and insights whose analysis would occupy the rest of his life. On his return Darwin settled in London, not far from Lyell's home. The two became friends, and Lyell helped Darwin's election to the Geological Society. Much later, Lyell would help Darwin with the 'delicate arrangement' that had to be worked out when Alfred Wallace, working independently, almost published a theory of evolution that would have trumped Darwin.

Over the following decades, Darwin used his *Beagle* data – and such input as Thomas Malthus's *Essay on the Principle of Population* – to establish his theory of natural selection: with variation from one generation to the next, followed by a relentless and lethal pruning by competition and predation, life and death worked together to shape species. Darwin had found a way in which a species could be shaped to fit its environment – not by divine intervention, not by mind, but through the steady, relentless working of natural law. Just like the Huttonian prescription for the Earth, it was a Newtonian scheme for life. For better or worse, Darwin transformed our view of our place in the universe. Humans too are not the outcome of a divine design, but simply products of the relentless workings of natural laws, just like rivers and mountains, beetles and whales.

For evolution to work, Darwin saw, he needed *time*: lots of it, time enough for the slow working of natural selection to turn a handful of windblown finches into the finely adapted varieties he had seen on Galapagos – time to turn a thing like a dog into a whale, an ape into a human.

Darwin's *Origin of Species* was dedicated to, among others, Charles Lyell. 'The great merit of [Lyell's] *Principles*,' he wrote, 'was that it altered the tone of one's mind.' Darwin knew little or nothing of Hutton: by Darwin's time, his work thoroughly assimilated and reworked by Lyell and others, Hutton had already become a figure of history. It was thanks to Hutton, however, that Darwin had time in abundance.

In 1856, when Darwin first showed his ideas to Lyell, the great geologist wasn't enthused. By now, Lyell had been defending his strict uniformitarianism for the best part of three decades. For a long time, it had been apparent that his extension of uniformity to living things was the shakiest of his tenets; the rocks showed that life was not static or recurring, as Lyell predicted with his flitting pterodactyls, but in some sense progressive: there were creatures in the rocks who had *not* recovered from their extinction. At fifty-nine, it isn't so easy to give up beliefs that have been cherished for a lifetime. However, Lyell could no longer defend his position, and he was forced to accept Darwin's natural selection.

But Lyell wasn't the only one with doubts about Darwin's synthesis – and still, six decades after his death, Hutton's ideas were to be challenged.

There is a 'Kelvin Room' at the home of the Royal Society of Edinburgh. It is decorated in a gaudy red flock wallpaper that reminds the Society's staff of an Indian restaurant. It contains a rather magnificent collage photograph of all the Fellows of the Society in 1902, arrayed as if they were brought together in a lecture room, like an intellectually weighty version of the Beatles' *Sergeant Pepper* cover. And there, front and centre, is Lord Kelvin himself, aged seventy-six, scowling like a particularly ferocious Old Testament prophet.

Lord Kelvin – born William Thomson – was the most formidable physicist of his generation. His name is remembered today for the Kelvin scale of temperature: absolute zero is the zero of his scale. And well might he scowl, for Kelvin hated evolution, finding Darwin's cold mechanical processes repugnant.

In December 1860, aged a mere thirty-six and laid up with a bad leg (broken in a curling accident), Kelvin, brooding on the origin of species, came up with a way to starve Darwin of the one resource his theory couldn't do without: Hutton's deep time.

How did the sun shine? It had been thought that it must be powered by a chemical burning – perhaps it was just a ball of coal. But Kelvin himself had shown that coal would burn out in a few thousand years.

Perhaps, it was proposed, the source of the sun's heat was gravity. The sun was held to be contracting. If you drop a rock down a mineshaft, gravitational energy is released, which shatters the rock and creates heat and noise. In the same way, the sun's infalling layers would release energy, converted to heat and light. But this couldn't last for ever. 'Within a finite period of time past,' Kelvin wrote, 'the Earth must have been and within a finite period of time to come must again be unfit for the habitation of man as at present constituted.' Kelvin's first calculation indicated that gravitational infall could keep the sun shining for a total lifetime of perhaps a hundred million years.

This is a long time – but Kelvin's results contradicted the sorts of dates the geologists and biologists were coming up with. Darwin himself had quoted over *twenty kilometres* as the total depth of British sedimentary rock: if present processes really were the key to the past, to have built up such a monstrous thickness must mean the past had to be long indeed. Just as Ussher's handful of millennia had in the end proved too constrictive, so now there just wasn't enough time even in Kelvin's hundred million years.

Darwin didn't have the physics or mathematics to counter Kelvin. But Kelvin's rapidly dying universe depressed him: 'Even personal annihilation sinks in my mind into insignificance compared with ... the certainty of the sun some day cooling & we all freezing. To think of the progress of millions of years, with every continent swarming with good & enlightened men all ending in this.' Still, Darwin went to his death convinced that the world would eventually be found to be 'rather older than Kelvin makes it'.

The crucial discoveries that would resolve this impasse were unassuming – even accidental.

By the end of the nineteenth century, physicists had grown puzzled by the phenomenon of phosphorescence: the way uranium salt crystals, for example, glowed in the dark. There had to be a new energy source *within* the uranium salt itself. By 1898, Marie and Pierre Curie had labelled this new energy radioactivity, and went on to show that a gram of radium could raise its own mass of water from freezing to boiling in an hour.

Suddenly, there was a new energy source to power the sun. The Earth, too, is heated continually by the disintegration of radioactive atoms in its interior. In 1903, the New Zealand physicist Ernest Rutherford was prepared to go public with the implications: 'The discovery of the radioactive elements ... thus increases the possible limit of the duration of life on this planet, and allows the time claimed by the geologist and biologist for the process of evolution.'

Kelvin himself was too old to back down: he was a great man, but he was stubborn. He always insisted that 'heavier than air flying machines are impossible' – though he died in 1907, four years after the Wright Brothers flew at Kitty Hawk.

Kelvin had at least made a start in applying quantitative physical principles to the *absolute* determination of Earth's age. Geological methods like stratigraphy could only tell you that one rock was older than another. To be able to work out how old a rock was in actual *years* was an extraordinarily exciting idea.

In studying the new radioactive materials, Rutherford quickly came to see that the way they decayed might lead to a new method for reaching Earth's deep past. The radioactivity of a material like uranium depended on the break-up of the nuclei of its atoms to give the nuclei of 'daughter products', such as helium or lead. But that meant, Rutherford saw, that if you had a sample of rock containing uranium ore, you could tell how old the rock was by comparing the proportion of the element itself to the daughter products.

Even Rutherford's first crude results were remarkable. The very first sample of pitchblende he tested showed an age of *seven* hundred million years, seven times Kelvin's notional age of the Earth. Rutherford's pioneering work was carried on by a young Englishman called Arthur Holmes, a physicist who had switched to geology. Holmes began a programme of extremely precise work which, by the 1930s, had pushed the minimum age of the Earth up into the *billions* of years.

The twentieth-century work on radioactivity provided a startling demonstration of Hutton's uniformitarianism. Even from the most ancient samples, the physicists produce devastatingly regular straight-line graphs – called 'isochrons' – which show beyond dispute the relentless and unchanging nature of the rate of radioactive decay. While these bits of rock were buried in the Earth, civilisations rose and fell, mountain ranges bloomed and disappeared like smoke. But to the rocks nothing mattered but the steady transformation of parent element to daughter, a steady nuclear ticking that continued unvarying and unperturbed, perhaps across billions of years.

But how old is Earth itself? Ours is an active planet; few rocks are left undisturbed for long. To determine the Earth's true age, what was needed was a piece of primordial material – the stuff of which Earth had been made in the first place.

American physicist Clair Patterson worked on the Manhattan Project of the Second World War, and would agonise for the rest of his life over the morality of the atomic bomb. To fix the age of the Earth, Patterson's academic supervisor suggested dating meteorites – bits of the primordial cloud from which the planets had formed, untouched through aeons until they fell to Earth. The procedure would be 'duck soup', said the supervisor. It turned out to be anything but. It took seven years before Patterson was confident that he had isolated his meteorite samples from lead contamination in Earth's environment. The level of contamination shocked him, and later in life Patterson would be instrumental in getting the USA's Clean Air Act on to the statute books.

Patterson finally succeeded in deriving a new date for the Earth: 4.55 billion years, give or take a few tens of millions of years. This date, published in 1956, has survived the half-century of study and confirmation since.

All the continents have ancient rocks: Asia, South America and Africa have all yielded samples more than 3.5 billion years old, while the oldest on the North American continent, from Canada's Slave Province, is more than four billion years old.

Perhaps the most astonishing rocks of all come from the Yilgarn Block in Western Australia. In January 2001 it was reported that a Yilgarn fragment had yielded a date of 4.4 billion years – just a hundred million years after the formation of Earth itself.

And the Yilgarn sample had crystallised from magma in contact with liquid water, and was embedded in quartzite, metamorphosed sandstone. Even on such a young Earth there were oceans, and on their shallow beds sediments gathered: then, as now.

Two centuries after Hutton's death, the triumph of deep time is complete.

21

'The bold outline traced with so masterly a hand'

Playfair, presciently, had understood how future generations would have to build on Hutton's theory. Previous thinkers, from Aristotle to Burnet, had tended to present their theories as unique, if not God-given inspirations; if they discussed their predecessors at all, it was only to dismiss them. But by 1800 a new idea had emerged: that human knowledge itself could evolve. A proposal like Hutton's was not a finished article, but could be taken up by a new generation, reworked and reshaped, corrected and modified, in the continuing search for a better understanding of the world. Playfair wrote, 'Ages may be required to fill up the bold outline which Dr Hutton has traced with so masterly a hand; to detach the parts more completely from the general mass; to adjust the size and position of the subordinate members; and to give to the whole piece the exact proportion and true colouring of nature.'

It has taken two centuries, and the overcoming of much intellectual resistance, but as Playfair hoped, James Hutton's 'bold outline' has indeed been completed and extended by later generations of workers, from Lyell to the present day. But it is all still founded on Hutton's basic insights.

Take the rock cycle, for instance. Hutton was the first to propose the constant exchange of material between Earth's surface and its interior.

At the surface of the Earth, weathering is relentless – just as Hutton observed – and even the hardest rock will, with time, erode away. The weathering can be chemical, as limestone is dissolved away by acid rainwater, or physical, as ice or salt crystals forming in cracks shatter rock surfaces. Even heat swings can fracture rocks: Hannibal used fire to break boulders on his way across the Alps.

As Hutton understood, a key product of weathering is soil.

Soil, composed of rock fragments and of decaying organic matter, is itself a complex system, embedded in the greater interacting systems of the Earth. Unlike Hutton, though, modern scientists would never describe the process of its creation as intentionally designed.

As weathering breaks down Earth's surface, the land is denuded through winds, rock falls and landslides, and through the action of rivers, tides and glaciers. Every year, some twenty thousand million tonnes of material are transported – passengers of water, wind or ice. All this debris is eventually deposited somewhere. A river will drop its heavier rocks and pebbles quickly, and when it reaches the ocean, coarser sands settle out close to the river mouth, then finer silts and clays are carried further out to sea. Thus as the land is constantly lowered, the sea bed gets ever thicker.

Hutton proposed that temperature and pressure, from the weight of overlying layers, combine to fuse this debris into rock. Mud and clay will turn to shale, sand to sandstone, gravel to conglomerate. Today we know that chemical action is also important. Under the weight of overlying layers, the mineral grains of a sediment are crushed together. As water is squeezed out, chemicals dissolved in it – calcite, quartz or iron oxides – can be deposited as cement to bind the grains together. Or the cementation can be caused simply by physical pressure; at points of contact the grains dissolve, and when they recrystallise they are locked together.

Organic matter can also be the source of sedimentary rocks. In the oceans, millions of microscopic creatures build their bodies from calcium carbonate dissolved in the sea water. When they die, they settle to the ocean floor, taking with them the minerals locked in their tiny bodies: chalk is four-fifths calcium carbonate. Coal is a special kind of sedimentary rock, formed from the remains of trees and other plants, laid down in stagnant lakes or swamps where the processes of decay are arrested.

Hutton was essentially right, too, that the internal heat engine of the Earth can drive the uplift of rocks.

We now know that tectonic movements, the collision of continents floating on great magmatic currents in the mantle, can thrust rocks up from deep inside the Earth's crust and pile them into mountains – the most spectacular examples today are the Himalayas and the Alps – which immediately begin to be eroded away in their turn. In this vast chthonic churning, rocks are dragged down into the ferocious cauldron of Earth's mantle, where they are melted and made anew. We can reconstruct these journeys using, in part, the experimental geology pioneered by Sir James Hall. In the Kokchetav Massif of Kazakhstan can be found metamorphic rocks containing micro-diamonds, formed from marine sediments probably transported more than a hundred kilometres deep, deeper even than the granite roots of mountains. And the isotopic composition of some basalts from Hawaii prove that they came from bits of ocean crust that had been carried nearly *half way down* into the Earth's interior, and exhumed once more. Just as Hutton wrote in his 1785 *Abstract*, 'The bottom of the ocean is to be made to change its place with relation to the centre of the Earth.'

James Hutton was the first to recognise the central importance of heat in the formation and operation of the Earth's features. He was the first to understand the importance of erosion, consolidation and uplift, and the interconnection of Earth's surface and underground forces. Hutton was the first to insist on the basic orderliness and constancy of geological processes through time: his uniformitarianism was the essential foundation of the orderly thinking that has remained the basis of geology ever since – even though, as geologists have learned of such processes as tectonic drift, glaciations and extraterrestrial impacts, they have had to develop their theories to incorporate catastrophic events. Today the debate is focused on the relative importance of fast and slow processes over geological time.

There have been many great geologists, but no figure before or since bequeathed a package of so many profound and integrated insights as James Hutton. And he was the first to construct a model of Earth's history containing its most essential feature: a vast and deep abyss of time.

But – to quote Stephen Jay Gould – though Hutton was a great thinker, he was not a modern thinker. And he has been hugely misunderstood.

*

Since his death, a heroic mythology has grown up around Hutton. This view has it that he was a fully modern scientist, proceeding by observation and fieldwork to deduction.

Lyell invented this Hutton as an ally in his war against the catastrophists, but it was Sir Archibald Geikie who cemented Hutton into the pantheon of progressive modern scientists. Geikie was himself a distinguished geologist and director-general of the British Geological Survey, who did much to revive the tradition of Scottish geology, which had foundered somewhat after Hutton's generation. In 1897, Geikie claimed that Hutton 'went far afield in search of facts ... For about thirty years, he never ceased to study the natural history of the globe ... In the whole of Hutton's doctrine, he vigorously guarded himself against the admission of any principle which could not be founded on observation. He made no assumptions. Every step in his deductions was based upon actual fact.' And so on.

The facts don't bear out this myth. Hutton clearly was a keen observer of the geological world. He spent thirty years studying formations and collecting specimens, and towards the end of his career he could read the land well enough to let it lead him straight to the kind of features he hoped to find. But still, his 1785 presentation had *preceded* the crucial evidence he needed: before 1785 he had seen only one unconvincing granite sample, and he had certainly seen no unconformities with the clarity of Jedburgh or Siccar Point.

And, like all of us, Hutton was a man of his time, immersed in his culture. In addition to 'scientific' logic and observation, his thinking was a rich mixture of metaphor and analogy and flights of fancy, much of which helped guide and shape his geology.

Some of Hutton's ideas, dismissed by subsequent generations, have resonance today.

Take his belief that the Earth was a kind of machine, a 'Living World', designed to sustain life. The 'Great Chain of Being' idea was commonplace in Hutton's day, but in later rationalist times it fell out of favour. As late as 1947, the commentator S. I. Tomkeieff, remarking on Hutton's geological philosophy, wrote that Hutton's imagery was an example of 'extreme and rather crude pictures of the Earth as a sort of superorganism [which] now seem fantastic'. Times change: now we can recognise that Hutton's deep intuition was actually a remarkable foreshadowing of another modern idea.

Today we understand that the Earth's surface – that is, its crust, the atmosphere, the water in the oceans and rivers and suspended in the air, and the biosphere, the world's great cargo of living things – is indeed a complex, interconnected system, constantly in flux under pressure of powerful forces. The two main driving mechanisms are the sun's radiant energy, which drives wind, rain and the other agents that attack and erode the Earth's surface, and Earth's internal engine, principally the movement of the great tectonic plates as driven by mantle convection currents, which create land relief by uplifting and buckling the land, as well as causing earthquakes and volcanoes. Just as Hutton intuited, these two forces are in a kind of balance, maintaining cycles of material.

And these tremendous cycles keep Earth habitable.

The astrophysicists have long wrestled with the 'Young Sun Paradox'. The sun, like all similar stars, is slowly brightening as it ages; in Earth's early history its power output was only some seventy per cent of its current value. But as far back as we can see, conditions on Earth have been equable. Yes, there have been intervals of glaciation, but on the whole, liquid water has been able to exist on Earth's surface for almost all of its history. Faced with a relentlessly brightening sun, some mechanism seems to have maintained the mean surface temperature of Earth in a range suitable for liquid water – in fact, for life.

The key turns out to be carbon dioxide, the notorious 'greenhouse gas' causing our current pulse of global warming: in the past it worked to trap the sun's heat, just as it does now. Carbon dioxide is injected into the air by outgassing from volcanoes and other tectonic phenomena, as well as from human industry. It is removed by weathering, as the gas combines chemically with surface rocks and so is drawn out of the atmosphere. The outgassing is more or less constant, but the weathering rate changes with temperature: it increases when the air heats up. So there is a feedback mechanism operating here, with the carbon dioxide concentration adjusting to the climate conditions, resulting in a 'homeostasis', a stable condition.

All this is just inanimate chemistry. But in the process of photosynthesis, plants ingest carbon dioxide – so the level of carbon dioxide in the air is intimately linked with life.

This, and a number of other biochemical and geochemical feedback cycles, led James Lovelock to formulate his 'Gaia hypothesis', proposing that life on a planetary scale has the ability actively to control its environment, rather than passively coping with changes. Lovelock's ideas were greeted by a predictable storm, but the records of the evenness of temperatures in the past, and similar data, seem unarguable.

Just as Hutton saw, our Earth is a 'living world' that, through its great cycles, has maintained conditions of equilibrium on its surface for billions of years, and will continue to do so for a long time to come. There is surely no mind involved, no intention. But to understand life – ourselves, and our future as part of the biosphere – we need generous and holistic thinking; we need to welcome back the spirit of Hutton's Enlightenment.

If Hutton were working today, the most controversial aspect of his theorising would certainly be his use of design arguments: his belief that the world is the way it is because God *intended* it so. But again, these ideas have a resonance today.

To many thinkers of Hutton's generation, the exquisite perfection of the living world seemed clear evidence of the work of God. 'Suppose I found a watch upon the ground,' wrote the theologian William Paley in 1802, 'and it should be inquired how the watch happened to be in that place ... When we come to inspect the watch, we perceive ... that its several parts are framed and out together *for a purpose*' (my emphasis). How could such an intricate thing as an eye or a wing have emerged from natural processes – how could it exist without a designing mind behind it?

Hutton actually did consider the hypothesis that it might have been chance that created the Earth and its cargo of life forms – that Paley's watch might indeed have emerged by sheer random luck. 'Was it the work of accident, or effect of occasional transaction, that by which the sea had covered our land? Or, was it the intention of that Mind which formed the matter of the globe, which imbued that matter with its active and its passive powers, and which placed it with so much wisdom among a numberless collection of bodies, all moving in a system?' It was surely more intellectually satisfying to believe that a Mind had created the exquisite order Hutton observed all around him. And after all, when Hutton was alive, nobody had any coherent alternative explanation: it was God, of one stripe or another, or nothing.

It would take Charles Darwin, a generation later, to show that randomness, through natural selection, could indeed produce the complexity of a living world, without the need for a Mind, divine or otherwise. In modern times Richard Dawkins has been prepared to put this even more bluntly: 'The universe we observe has precisely the properties we should expect if there is, at bottom, no design, no purpose, no evil and no good, nothing but blind, pitiless indifference.' This is at once the beauty and the terror of our new world view, and perhaps it is no surprise that many find it difficult to come to terms with such a cold vision.

Remarkably, however, design principles have made a comeback in recent years, in a rather different form.

It has become apparent that it required a long chain of cosmic coincidences to produce a universe, a world, in which creatures like us can exist to view it. For example, our biochemistry is based on carbon. But at one time there was *no* carbon in the universe. Only hydrogen and helium, and traces of other light elements, emerged from the Big Bang. The heavier elements, including carbon, were cooked in stars, and then scattered in supernova explosions, to become available for making cells and bones, wings and eyes and people.

But the formation of carbon in stars, by a complex chain of fusion processes, depends on a precise coordination between different physical constants that, if it were even slightly off, would fail. It is as if the universe has been fine-tuned to produce us (or at least carbon!). The strings of 'coincidences' required to produce us are so long that some have been tempted to speculate there may be a purpose to it all. Fred Hoyle – the British astrophysicist who unravelled the production of carbon in stars – said in 1959, 'I do not believe that any scientist who examined the evidence would fail to draw the inference that the laws of nuclear physics have been *deliberately designed* with regard to the consequences they produce inside the stars' (my emphasis).

Other thinkers, including physicists Frank Tipler and John Barrow, have gone further, and developed this modern design-oriented thinking into a 'cosmological anthropic principle', exploring the possibilities that the universe was in some sense designed to produce human beings. Or perhaps the universe has itself been shaped by evolution, the product of natural selection among some meta-cosmic population.

My point here is not to say that design arguments, for example, are right or wrong – though I certainly don't condone their misuse by modern creationists. But as Hoyle, Tipler and Barrow show, such ideas can be fruitful and suggestive of further research – just as Harvey's belief about a design in the blood's circulation system led to a prediction about the existence of capillaries, and Hutton's design arguments about Earth's replenishment led him to look for, and eventually find, a renewal mechanism in geological uplift.

Does it matter how Hutton came up with his ideas?

Yes, I believe it does. To understand how scientific and other insights come about, we have to reconstruct the thinking of figures like Hutton, as accurately as we can. After all, a new paradigm can be constructed only by a thinker educated in the old paradigm: you have to be wrong before you can be right. And if we choose to inspect the past solely through the prejudices of the present, we will fail fully to appreciate the richness of human thought. We will lose other ways of thinking that, just like Hutton's ideas and metaphors, may have much to offer us as we confront the new problems of today.

We will always need vision. The Princeton physicist Freeman Dyson has looked even beyond the end of the Earth, to sketch the far future of life in an expanding universe. Dyson's work was based on the same thinking as Hutton's, that physical laws will continue to work as we understand them into the indefinite future: if the present is the key to the past, it is also the key to the future. And 'no matter how far we go into the future,' said Dyson, 'there will always be new things happening, new information coming in, new worlds to explore, a constantly expanding domain of life, consciousness, and memory.'

As James Hutton firmly believed, the future is full of hope.

EPILOGUE

Hutton was buried in Greyfriars Kirkyard in Edinburgh, not far from the grave of his friend Joseph Black. The Kirkyard is an island of surprising peace, surrounded as it is by the bustle and traffic of the Old Town. Hutton's grave was unmarked, and nobody knows quite where it is, but in 1947, on the 150th anniversary of Hutton's death, the Lord Provost unveiled a plaque to 'James Hutton MD FRSE: Founder of Modern Geology'.

After Hutton's death it was quite a surprise to everyone when James Hutton Junior – by then a man in his forties – turned up in Edinburgh, seeking financial help. None of Hutton's friends had known anything of James Jr's existence – not even Black. But Black and Watt knew the mother, for Black wrote to Watt that 'he is not like [Hutton] in the face, having more the features of his mother.'

As Hutton had died unmarried and intestate – James Jr was not a legitimate heir – his property passed to his only surviving sister, Isabella. Isabella, a spinster herself, remained living in the house on St John's Hill. In 1810, she sold the farm at Slighhouses to one Charles Douglas, and on her death in 1821 she left Nether Monynut to Hutton's grandchildren by James Jr: Edington, Margaret and Jane Smeaton Hutton. Two years after that, the Smeaton Huttons sold the farm on for £1,580.

As for Hutton's house, that would survive until the 1960s, when a disastrous programme of redevelopment obliterated that part of Edinburgh. In 1997, conferences were held to celebrate the bicentenaries of Hutton's death, and Lyell's birth. To mark the occasion, money was raised by international subscription to establish a kind of memorial garden at the site of Hutton's house. The garden is unprepossessing, sandwiched between a multi-storey car park and the ugly tenement blocks that replaced the architecture of Hutton's era. But there is a plaque, mounted on a bit of Triassic sandstone, bearing a cartoon of Hutton and an inscription of his 'no vestige ... no prospect' quotation, and there are boulders from Glen Tilt and elsewhere. And the Salisbury Crags loom over it all, as they always did. Humans and their petty doings come and go, but the geology endures.

James Davie, Hutton's partner in the sal ammoniac business, died not long after Hutton himself. By now their manufacturing process was becoming old-fashioned. The business seems to have been quickly wound up, for soon after Davie's death Edinburgh soot was being sent to a works in Yorkshire.

By 1800, when James Watt's partnership with Boulton ended and the patent on the separate-condenser system expired, 451 machines had been built to Watt's designs. Britain's miners were working up to three hundred metres deep, and the mines produced a million tonnes of coal a year. Watt had turned the steam engine from Newcomen's clanking and inefficient water pump into a universal source of power for industry and transport. Watt himself suffered indifferent health until his death in 1819. In his old age he was honoured. His memorial in Westminster Abbey acclaims him as 'among the most illustrious followers of science and the real benefactors of the world'.

Richard Kirwan too gathered more honours. Life President of his precious Royal Irish Academy, he was also appointed president of the Dublin Library Society and inspector-general of His Majesty's Mines in Ireland, and he was elected to many foreign academies including Berlin, Stockholm, and Philadelphia, the capital of newly independent America. There was even a Kirwanian Society of Dublin founded as a tribute to him and to spread his ideas.

Kirwan himself became still more eccentric. He cultivated a habit of patrolling the grounds of his estate with wolfhounds and greyhounds: he had developed an attachment to large dogs, since an Irish wolfhound had rescued him from an attack of six wild boars. Sometimes he would be accompanied by a huge eagle that would perch on his shoulder; he had trained the eagle himself and it had become devoted to him. One day, as it swooped down towards Kirwan's shoulder, a friend thought the eagle was going to attack him and shot it.

When Kirwan died in 1812, at the age of seventy-nine, his personal copy of Hutton's *Theory* would be found with many of its pages uncut. Richard Kirwan had become so obsessed with Hutton's theories that he had written a whole book about them, but he hadn't bothered to read through Hutton's *Theory*: Kirwan *knew* Hutton was wrong without even having to check.

Sir James Hall died in 1832, aged seventy-one. He was buried in the Dunglass Collegiate Church on his family's estate.

Today the Church is maintained as a monument; you can find it easily by following the brown tourist-information signs from the A1. The estate is a beautiful, open place, in harmony with the rolling countryside, with a view of the sea that is somewhat spoiled by the squat monolith of Torness. The Church itself, dedicated to the Virgin Mary, was completed in 1450. It is a small, pretty sandstone building, with a massive square tower at the centre, a vaulted choir and nave, and a stone slab roof. It was successfully defended against Henry VIII's army in 1544, but in the seventeenth century it was converted into a stable; its east wall was pulled down to let in the animals.

The south transept is the 'burial aisle' of the Halls. Carved stone memorial plaques are set in the walls, and I spent a happy morning sitting in the transept reconstructing Hall's family history, running from his father Sir John, born 1711, to his grandson, died 1876. The family's story weaves in and out of British history. Sir James Hall married Helen, who bore him six children. His daughter Magdalene married a Colonel William de Lancey who was killed at Waterloo, and one of Hall's grandsons served in the siege of Sebastopol.

Sir James Hall's own plaque records that he was 'distinguished among the eminent men of an enquiring age, not less by the originality, boldness and accuracy of his speculations than by the ingenuity and resolute perseverance with which he substantiated various important theoretical views in his favourite science of geology by a series of brilliant and convincing experimental researches.' Not a bad epitaph.

The Royal Society of Edinburgh has continued to flourish. Hall was president from 1812 to 1820, and Playfair served as general secretary from 1798 to 1819, during which period he was described as 'the life and soul of that institution'. The Society held its meetings in the cramped quarters of the University Library until 1810, when it moved to a house in George Street, and then in 1826 to the Royal Institution. But it was only a tenant, and after eighty years it was startled to be ejected so that the Royal Institution could become an art gallery. After much lobbying by Kelvin and others, Parliament granted the Society a new home at its present residence in George Street. There, a huge portrait of an elderly James Hall, ferociously reading an unidentifiable book, looms over the entrance lobby, and there is a plaster bust of Hutton himself, in neo-classical mode.

A key purpose of the Society was always to provide an opportunity for scientists and scholars of different specialisms to mix. The clubbable Hutton, Smith, Ferguson, Black and others hardly needed any encouragement, but a century later 'it was Kelvin moving eagerly on the soft carpet, and putting his gyroscopes through their dynamical drill ... or Lister quaffing a glass of milk which had stood for weeks under a light stopper which no germs could creep through'.

Today, through such activities as lectures for schoolchildren and seminars, the Society continues to pursue its goal of 'the cultivation of every branch of science, erudition and taste'. Among the Society's notable achievements was the 'Challenger' expedition, a mighty oceanographic feat covering 1,000 days and 69,000 nautical miles. But Hutton's 1788 paper on his theory of the Earth is still listed as one of the top five published by the Society in its history of more than two hundred years.

The Oyster Club, though, did not survive its founders.

On Hutton's death, Isabella gave his precious rock collection to Joseph Black, who was by then old and infirm himself. So he offered the collection to the Royal Society of Edinburgh, on the conditions that it be kept together as a Huttonian collection, that trustees should be appointed, and that the specimens should be properly labelled and catalogued.

The Society, under its Charter, had to deposit any collections where they could be made available to scholars and the general public. In practice that meant Edinburgh University's Museum of Natural History – where the specimens came into the care of the Reverend John Walker, professor of natural history. From the time of Hutton's 1785 presentation, Walker had been vehemently opposed to Hutton the system-builder. Not only that, Walker would soon be succeeded in his post by his nephew Robert Jameson, Werner's most influential British follower. In summer 1797, four months after Hutton's death, Jameson visited Kirwan in Dublin.

Jameson was a young man who had inherited a museum left in a muddle by an ailing Walker, and it may be that he failed to understand how Hutton had used his 'hand specimens' to establish his arguments, and therefore simply failed to grasp the importance of the collection – but it's hard to avoid the conclusion that he was motivated by a Wernerian prejudice. Jameson didn't even try to catalogue the collection. He wouldn't display it to the public, or even to the Fellows of the Royal Society. When Playfair and two other trustees finally forced their way in to see the collection, they found 'many of the most important specimens were missing'.

After that the collection was steadily lost: by 1835 it was being kept in four small cases. What was left must have been only a fraction of Hutton's legacy – the great boulders from Glen Tilt and Arran had certainly gone. In 1855 the entire contents of the Natural History Museum were transferred to a new museum, now known as the Royal Scottish Museum. But the new museum's catalogue does not mention Hutton's material. The last remnants of the collection must have been re-labelled, dispersed or discarded before the move; even if any of Hutton's specimens did finish up in the Royal Museum, it would be impossible to identify them.

Thus Hutton's precious collection was destroyed, his 'God's Library' broken up at the hands of his intellectual opponents. None of Hutton's own rocks would survive to be shown in his 1997 rock garden on St John's Hill.

As for Hutton's manuscript on agriculture, Playfair dutifully read it through, but it appears never to have been published. Bound up in two volumes, the manuscript disappeared from view for sixty years. It came at last into the hands of James Melvin, vice-president of the Edinburgh Geological Society, and in 1887 he gave it to that Society. Since 1949 the manuscript has been on permanent loan to the Royal Society of Edinburgh.

The rest of Hutton's archive was not spared. As was the practice of the time, his executors destroyed much material we would now treasure, including most of his manuscripts and his letters. Even Hutton's field notebooks were burned.

Handwritten drafts of some of the chapters that would have been the backbone of the two later volumes of Hutton's *Theory of the Earth* did pass into Playfair's hands from Lord Webb Seymour, who had made the return trip to Glen Tilt with Playfair in 1807. Playfair tried to publish the material, but there were difficulties with obtaining the engravings. When John Clerk of Eldin died in 1812, the reconstruction of the drawings became impossible.

The manuscript eventually passed to Leonard Horner. Horner was the younger brother of Francis, who had mused on cycles of history during Hutton's 1785 presentation. Horner became a mineralogist, rising to become president of the Geological Society; Charles Lyell married Horner's daughter, Mary. In 1856, Horner donated some of the Hutton manuscripts to the safekeeping of the library of the Geological Society at Burlington House in London. These bound manuscripts were set on a shelf alongside the two printed volumes of the *Theory*.

There they were lost, and forgotten.

In the 1890s, a Canadian geologist called Frank D. Adams came to the library. Adams asked to see the two published volumes of the *Theory* – and was startled to be handed in addition a shabby bound manuscript. The library assistant thought Adams might find something of interest in 'this old thing'. It was the Hutton essays.

In June 2002, I visited Burlington House to view Hutton's manuscript. Burlington House is just off Piccadilly, London. Like many of Britain's great institutions, there is a vaguely museum-like feel to the place: I read the Hutton manuscript in the Upper Library, a multistorey atrium with a glass roof and pillars, where chandelier light glints from computer terminals, and clocks chime the hours melodically.

The manuscript itself is unprepossessing, a little more than two hundred quarto-sized pages hard-bound in pale brown cloth. Evidently the manuscript was left lying around for some time, for the first page is soiled and torn, and had to be repaired by sticking a leaf to its back. The pages are unlined and hand-written on one side of the paper. It felt strange and a little scary to handle a manuscript so *old*. It seems remarkable it has survived as well as it has; the paper is largely free of yellowing and unsoiled.

Horner treasured the fragile pages that had come into his possession. In a pasted-in frontispiece, beneath his coat of arms, he wrote by hand, in a note dated November 1856, of how he came into the possession of 'This ms. volume (part of a series) of Dr Hutton's, with some additions in his own hand ... I give it to the Geological Society, to be preserved in the Library, as an interesting document in the History of the Science.'

What we actually have here is a little mysterious. As Horner notes, this volume is one of a series – but it is the only one we have. The pages are numbered, by hand, from p. 139 to p. 346, and we have six chapters numbered from Four to Nine. Chapters Four, Five and Nine are essentially field reports of Hutton's trips to Glen Tilt in 1785, Cairnsmore in 1786, and Arran in 1787. Chapters Six to Eight are write-ups of Hutton's reading of a new volume by de Saussure about the Alps published in 1786, of a book on the Pyrenees published in 1781, and a book on Calabria published in 1784.

When were these pieces written? The fieldtrip chapters are written in the form of diaries. In the Arran chapter, Hutton says he fulfilled his ambition to study Arran 'this summer 1787' (p. 296). But elsewhere in the same essay, he refers to letters by one Abraham Mills not published in the *Philosophical Transactions of the Royal Society* until 1790.

My own impression is that this manuscript was indeed written out after the drafting of Volumes One and Two of the *Theory*. Hutton had his fieldwork essays and notes on his reading to hand; he read through this material to his amanuensis somewhat hastily, with added observations. Perhaps Webb, Playfair or Horner knew what became of the rest of the 'series' of manuscript chapters, bound or otherwise, that would have made up Volumes Three and Four.

Even the material we have is sadly incomplete. There are references to plates now lost. Of one lost drawing by Clerk of a granite intrusion they saw near Crieff in 1786 Hutton said, '[it] will convince the most sceptical with regard to this doctrine of the transfusion of granite.'

For me, the main value of holding this manuscript was the feeling it gave that I was in the presence of Hutton himself. The manuscript has been written out in a neat, flowing copperplate by an amanuensis (perhaps Isabella) who took down Hutton's dictation, evidently for hours on end. But here and there Hutton has made his own hand corrections, striking out lines or adding notes in a bolder, blotchier, more uneven hand. The prose is verbose, the corrections few and far between; evidently this is a work dictated in haste.

The field reports are in places bright and vivid, and Hutton's personal observations are sometimes delightful: of Glen Rosa on Arran he says, 'in this dreary glen is to be found a charming picture of nature in decay, or of lofty mountains going into ruin, apparently without a purpose' (p. 311). He was friendly and inclusive: on Arran, having happily observed a junction between granite and schistus, he made an expedition to another location called Glen Shant as 'I wished to give Mr Clerk the same satisfaction' (p. 314).

Hutton's *only* surviving geological section shows how this junction proceeds 'in great steps'. It is just a little square figure drawn on p. 315, perhaps a centimetre across, with a staircase line dividing the schistus (horizontal bands) from the granite (dots).

Perhaps most touching of all is a footnote to p. 233, concerning Alpine structures, which Hutton wrote out on a scrap of paper that has been stuck to the back of p. 208. Horner has noted, 'This is most probably the handwriting of Dr Hutton. See the back of the paper.' And indeed, if you look carefully, you can make out an address: 'Doctor Hutton – St John's Hill'.

The unconformity at Siccar Point is a contact between rocks now thought to have been formed in the Silurian period, perhaps 420 million years ago, and the late Devonian, some 360 million years ago. So the rock types you see exposed under the cliffs are separated in time by some eighty million years – and even that gap is more than *ten thousand times* longer than all of Bishop Ussher's history.

As the shallow seas of the dinosaur ages flooded the Earth, great thicknesses of chalk were laid down over the unconformity, hiding it like an exhibit in some great buried museum: the geologists say it was a 'hidden landscape'. The chalk layers themselves took perhaps 150 million years to form. A mere two million years ago, the glaciers sliced away the softer surface rocks to expose the unconformity to the air, and Hutton's view. The unconformity is still there today, of course. But everything exposed to the air is subject to erosion. In a few more millions of years the unconformity will itself be gone, greywackes, sandstone, strata and all.

It is hard for us now to grasp the cosiness of Bishop Ussher's timescale. The whole history of Earth could be measured in a few human generations, and the mummies of Egypt had been buried in an age halfway back to creation itself. It was time built on a human scale, comprehensible and brief; and we humans were secure within its scriptural walls.

Hutton knocked all that away.

The abyss of time defies the imagination. We would expect a photograph of a landscape taken today to look essentially the same as one taken a hundred years ago. But over historical time we know that rivers have diverted from their courses, and estuaries have silted up, stranding many former sea ports deep in the interior of the land. On longer time scales, Earth's surface billows restlessly. The land surface of North America is washing away at the rate of a few hundredths of a millimetre each year – while the Niagara Falls, cut back at the rate of a metre a year, are positively evaporating. If you could fast-forward your perception of time, so that a million years passed in a few minutes, you would see mountains bloom and fade, as transient as flowers.

We are almost at the mid-point of Earth's lifespan of some ten billion years. Hutton said that in the rocky biography of Earth we would find 'no vestige of a beginning, no prospect of an end'. And he was right: the oldest rocks to be found on Earth's surface, survivors of more than four billion years of tectonic processing, are essentially the same as those being formed today. It is astrophysical events, events beyond Earth, that govern the beginning and the end of the world: Earth's birth from a cloud of jostling rocks, and its fiery death when the dying sun swells to a red giant. Modern humans have existed for perhaps a hundred thousand years, a fleeting instant compared to Earth's titanic biography, like a single second at noon set against a full day. Our knowledge of our brevity, and the awesome expanses of time around us, are Hutton's astonishing and terrifying gift.

Today, living in Hutton's shadow, we take an ancient Earth for granted. We can scarcely imagine the drama of those moments when James Hutton saw limitless time in a few broken rocks and a handful of soil.

ACKNOWLEDGEMENTS

Thanks to my agent Robert Kirby of PFD, London, for a bright idea. I'm grateful to my brother, Dr Anthony Baxter FFPHM, for his speculations on the cause of Hutton's death. Mr Bill Veitch of Jedburgh, Roxburghshire, was very kind in allowing me to visit Hutton's unconformity by the banks of the Jed, which is within his property. I'm grateful to Andrew Mussell and Wendy Cawthorne of the Geological Society of London for allowing me to view Hutton's geological manuscript. Ms Vicki Ingpen of the Royal Society of Edinburgh was exceptionally generous in assisting me with my research.

FURTHER READING

Source Material and General References

James Hutton, 'Theory of the Earth; or an Investigation of the Laws Observable in the Composition, Dissolution, and Restoration of Land upon the Globe', *Transactions of the Royal Society of Edinburgh*, vol. 1, pp. 209–305 (1788). This paper, first read by Hutton to the Royal Society of Edinburgh in 1785, was the first publication of his theory of Earth's cycles and age. The paper is in the public domain and is available, for example, at www.mala.bc.ca/~johnstoi/essays/hutton.html. The printed form has some modifications from Hutton's original 1785 oral presentation, but it is the best account we have of what was said.

James Hutton, *Theory of the Earth with Proofs and Illustrations* (Edinburgh, William Creech, 1795). This is Hutton's full explication of his theories. Only two volumes of a projected four were published in Hutton's lifetime.

James Hutton, *Geological MSS of Dr Hutton*, Geological Society of London archive holding LDGSL 753. Portions of the 'lost' volumes of Hutton's *Theory of the Earth*.

Dennis R. Dean, *James Hutton in the Field and in the Study* (New York: Scholars' Facsimiles and Reprints, 1997). Published for the bicentenary of Hutton's death, a handsome facsimile edition of Geikie's 1899 edition of the above, with illustrations and manuscript pages.

John Playfair, 'Account of the Late Dr James Hutton', *Transactions of the Royal Society of Edinburgh*, vol. 5, part 3, pp. 39–99 (1805). This is the primary source for biographical information on Hutton, a brief account prepared after his death by his close friend John Playfair and read to the Royal Society of Edinburgh on 10 January 1803. Playfair's aim, however, was to present Hutton and his work and thinking in what he saw as the best light; he was an evangelist for Hutton, and as history his writings have to be taken with a pinch of salt.

Dennis R. Dean, *James Hutton and the History of Geology* (New York: Cornell University Press, 1992). Detailed academic work on Hutton's writings and their reception by and influence on geologists.

Stephen Jay Gould, *Time's Arrow, Time's Cycle: Myth and Metaphor in the Discovery of Geological Time* (Harvard, Mass., 1987). A fine exploration of Hutton's scientific thinking, focusing on his use of design arguments – though unnecessarily harsh on Hutton's achievements as a field scientist. Also contains an interesting study of Burnet, and of Lyell's uniformitarianism.

Various authors, *Proceedings of the Royal Society of Edinburgh*, B vol. 63, pp. 351–400. This is the publication of papers from a conference held in 1947 to commemorate the 150th anniversary of Hutton's death. It includes 'James Hutton, Founder of Modern Geology' by E. B. Bailey, 'Hutton on Arran' by G. W. Tyrrell, and 'James Hutton and the Philosophy of Geology' by S. I. Tomkeieff.

G. Y. Craig and J. H. Hull (eds.), *James Hutton – Present and Future* (London: Geological Society, 1999). This volume is based on papers presented at a conference held in Edinburgh in 1997 to celebrate the bicentenary of Hutton's death. It includes useful papers relating to a modern updating of Hutton's ideas by P. Wyllie on Hall's experimental geology, W. Schreyer on the metamorphosis of rocks, and A. Watson on the Gaia hypothesis.

Prologue

Stephen Baxter, *Moonseed* (London: HarperCollins, 1998). A geological disaster story set in Edinburgh.

Gregory Benford, *Deep Time* (New York: Avon, 1999). A meditation on the modern view of time and mankind's attempts to challenge its immensity, from pyramids to interstellar spacecraft.

Donald McIntyre and Alan McKirdy, *James Hutton, The Founder of Modern Geology* (Edinburgh: HMSO, 1997). A brief popular biography of Hutton, focusing on the geology, but portraying Hutton retrospectively as a modern scientist.

One: Deposition

Iain Gordon Brown and Hugh Cheape, *Witness to Rebellion: John Maclean's Journal of the 'Forty-five and the Penicuik Drawings* (East Linton: Tuckwell Press, 1995). The war diary of one of Bonnie Prince Charlie's officers, and caricatures of figures of the Jacobite rising from the Clerk family collection at Penicuik.

Archibald Clow, 'Dr James Hutton and the Manufacture of Sal Ammoniac' *Nature*, vol. 159, pp. 425–27 (1947). How Hutton and Davie made a profit from Edinburgh soot.

Martin Gorst, *Aeons: The Search for the Beginning of Time* (London: Fourth Estate, 2001). Entertaining survey of efforts to determine the age of Earth and the universe, including biographical material on James Ussher.

Stephen Jay Gould, *Hen's Teeth and Horse's Toes* (London: W. W. Norton, 1983). This collection of essays contains illuminating pieces on Steno, Cuvier and Hutton himself.

Jean Jones, 'James Hutton's Agricultural Research and Life as a Farmer', *Annals of Science*, vol. 42, pp. 573–601 (1985). A useful biographical survey of Hutton's farming years, drawing on letters and an unpublished manuscript.

John Prebble, *The Lion in the North* (London: Penguin, 1981). One of this journalist and author's several excellent and vivid works on Scottish history.

Two: Consolidation

Neil Campbell and R. Martin Smellie, *The Royal Society of Edinburgh (1783–1983)* (Edinburgh: Royal Society of Edinburgh, 1983). A bicentenary celebration and history of the Society.

V. A. Eyles, 'Some Geological Correspondence of James Hutton', *Annals Of Science*, vol. 7, pp. 316–39 (1951). Some of Hutton's rare surviving correspondence, from about 1770.

Adam Ferguson, 'Minutes of the Life and Character of Joseph Black MD', *Transactions of the Royal Society of Edinburgh*, vol. 5, part 3, pp. 101–17 (1805). A biography of one of Hutton's friends by another, with illuminating insights into their relationships.

Patsy A. Gerstner, 'The Reaction to James Hutton's Use of Heat as a Geological Agent', *British Journal for the History of Science*, vol. 3, pp. 353–62 (1971). Exploration of Hutton's idiosyncratic views on the nature of heat.

Arthur Herman, *The Scottish Enlightenment: The Scots' Invention of the Modern World* (London: Fourth Estate, 2001). Enthusiastic if uncritical celebration of the Scottish Enlightenment and its impact.

James Hutton, 'A note on Adam Smith's death', *Transactions of the Royal Society of Edinburgh*, vol. 3, p. 131n.

Jean Jones, Hugh Torrens and Eric Robinson, 'The Correspondence between James Hutton (1726–1797) and James Watt (1736–1819) with Two Letters from Hutton to George Clerk-Maxwell (1715–1784)', *Annals of Science*, vol. 51, pp. 637–53 (1994), and vol. 52, p. 357–82 (1995). These letters give a rare glimpse of Hutton's private life and his fieldwork – especially his geological tour of 1774.

Donald B. McIntyre, 'James Hutton's Edinburgh: The Historical, Social, and Political Background', *Earth Sciences History*, vol. 16, pp. 100–57 (1997). Informative survey of Hutton's Edinburgh years.

J. E. O'Rourke, 'A Comparison of James Hutton's *Principles of Knowledge* and *Theory of the Earth*', *Isis*, vol. 69, pp. 5–20 (1978). An exploration of Hutton's theory of knowledge and how it applied to his geological thinking.

Stephanie Pain, 'The Flute-maker's Fiddle', *New Scientist*, 9 March 2002. Young James Watt's possible forgeries of flutes.

David Stevenson, *The Beggar's Benison: Sex Clubs of Enlightenment Scotland and their Rituals* (East Linton: Tuckwell Press, 2001). Analysis of the sexual underside of the Scottish Enlightenment.

Three: Uplift

John Barrow and Frank Tipler, *The Anthropic Cosmological*

Principle (Oxford: Oxford University Press, 1986). Sets out a modern version of the old arguments that the universe has in some sense been created to support man, and includes historical material on design arguments.

Stephen Baxter, *Deep Future* (London: Gollancz, 2001). A sketch of the far future of humanity in an expanding universe.

Richard Dawkins, *The Blind Watchmaker* (London: Longman, 1986). Dawkins' impassioned defence of Darwinism.

James Lawrence Powell, *Mysteries of Terra Firma* (New York: Free Press, 2001). Accessible account of the development of geology in the twentieth century.

James Lovelock, *Gaia: A New Look at Life on Earth* (Oxford: Oxford University Press, 1979). Lovelock's statement of his 'living world' hypothesis.

Peter Ward, *The End of Evolution* (London: Weidenfeld and Nicolson, 1995). A survey of mass extinctions by an expert in the field.

A. N. Wilson, *God's Funeral* (London: John Murray, 1999). An account of the crisis of faith in nineteenth-century Britain, partly predicated by geology.

Simon Winchester, *The Map That Changed the World: The Tale of William Smith and the Birth of a Science* (London: Viking, 2001). Informative if hagiographic study of the founder of stratigraphy.

Epilogue

C. D. Waterston, *Collection in Context: The Museum of the Royal Society of Edinburgh and the Inception of a National Museum for Scotland* (Edinburgh: NMS Publishing, 1997). Contains information on the fate of Hutton's rock collection.

INDEX

ABOUT THE AUTHOR

Stephen Baxter is the critically acclaimed author of *The Time Ships*, *Deep Future*, *Evolution*, *Moonseed*, and *Voyage*. With Arthur C. Clarke, he wrote *The Light of Other Days*. Born in 1957, he was raised in Liverpool and has a degree in mathematics from Cambridge University and a Ph.D. in aeroengineering from the University of Southampton. He has won the John W. Campbell Award, the Philip K. Dick Award, and other honors. He lives in Buckinghamshire, England.